# Migration, Tourism and Social Sustainability

The distinctions between tourism and migration are increasingly blurred. Tourism often drives various forms of mobility, and an international workforce is essential to maintaining functioning tourism economies. This book explores intersections of tourism and migration, considering their relationships with and impacts on social sustainability. The chapters explore in a variety of contexts how the planning, development and governance of tourism affects the sustainability of communities, which consequently influences attitudes towards migrants and tourists. They also consider how migrant-local connections may evolve, creating opportunities for positive, symbiotic co-existence or intergroup tensions and exploitative relationships. The book paves the way for future work examining new forms and interactions between migration and tourism that contribute to social sustainability.

This book will be of great value to students, academics, and researchers interested in tourism, geography, migration/diaspora studies and sociology. The chapters in this book were originally published as a special issue of *Tourism Geographies*.

**Jaeyeon Choe** is Lecturer at Glasgow School for Business and Society, Glasgow Caledonian University, UK and Visiting Professor at the School of Hospitality and Tourism at Hue University, Vietnam. Her primary research areas include the well-being of marginalised communities, social sustainability of destinations, and wellness/spiritual tourism in Southeast Asia.

**Peter Lugosi** is Professor of Culture and Organisation at the Oxford Brookes Business School and the Centre for Business, Society and Global Challenges at Oxford Brookes University. He has researched and published on a wide range of subjects including migrants' adaptation, labour market transition and entrepreneurship.

# Migration, Tourism and Social Sustainability

*Edited by*
**Jaeyeon Choe and Peter Lugosi**

Routledge
Taylor & Francis Group
LONDON AND NEW YORK

First published 2023
by Routledge
4 Park Square, Milton Park, Abingdon, Oxon OX14 4RN

and by Routledge
605 Third Avenue, New York, NY 10158

*Routledge is an imprint of the Taylor & Francis Group, an informa business*



*British Library Cataloguing in Publication Data*
A catalogue record for this book is available from the British Library

ISBN13: 978-1-032-41480-5 (hbk)
ISBN13: 978-1-032-41481-2 (pbk)
ISBN13: 978-1-003-35828-2 (ebk)

DOI: 10.4324/9781003358282

Typeset in Myriad Pro
by Newgen Publishing UK

**Publisher's Note**
The publisher accepts responsibility for any inconsistencies that may have arisen during the conversion of this book from journal articles to book chapters, namely the inclusion of journal terminology.

# Contents


| | | |
|---|---|---|
| | *Citation Information* | vii |
| | *Notes on Contributors* | ix |
| | Introduction: Migration, tourism and social sustainability<br>*Jaeyeon Choe and Peter Lugosi* | 1 |
| 1 | Who is the city for? Overtourism, lifestyle migration and social sustainability<br>*Jaime Jover and Ibán Díaz-Parra* | 9 |
| 2 | Zoning for world heritage sites: Dual dilemmas in development and demographics<br>*Thomas E. Jones, Huong T. Bui and Katsuhiro Ando* | 33 |
| 3 | Perceptions of and interactions between locals, migrants, and tourists in South Tyrol<br>*Anja Marcher, Ingrid Kofler, Elisa Innerhofer and Harald Pechlaner* | 56 |
| 4 | Local gastronomy, transnational labour: Farm-to-table tourism and migrant agricultural workers in Niagara-on-the-Lake, Canada<br>*Kristin Lozanski and Kayla Baumgartner* | 73 |
| 5 | Q method finds anti-refugee sentiments on Yemeni migration to Jeju<br>*Armigon Akhmedov, William Cannon Hunter and Jeong-Gil Choi* | 96 |
| 6 | "Traditional Mexican Midwifery" tourism excludes indigenous "others" and threatens sustainability<br>*Rosalynn A. Vega* | 117 |
| 7 | Exclave accessibility and cross-border travel: The pene-exclave of Ceuta, Spain<br>*Ioulia Poulaki, Andreas Papatheodorou, Alexandros Panagiotopoulos and Sotiroula Liasidou* | 141 |

8 Labour migration and tourism mobilities: Time to bring sustainability into the debate 166
*Noel B. Salazar*

*Index* 177

# Citation Information

The chapters in this book were originally published in the journal *Tourism Geographies*, volume 24, issue 1 (2022). When citing this material, please use the original page numbering for each article, as follows:

**Introduction**

*Migration, tourism and social sustainability*
Jaeyeon Choe and Peter Lugosi
*Tourism Geographies*, volume 24, issue 1 (2022), pp. 1–8

**Chapter 1**

*Who is the city for? Overtourism, lifestyle migration and social sustainability*
Jaime Jover and Ibán Díaz-Parra
*Tourism Geographies*, volume 24, issue 1 (2022), pp. 9–32

**Chapter 2**

*Zoning for world heritage sites: Dual dilemmas in development and demographics*
Thomas E. Jones, Huong T. Bui and Katsuhiro Ando
*Tourism Geographies*, volume 24, issue 1 (2022), pp. 33–55

**Chapter 3**

*Perceptions of and interactions between locals, migrants, and tourists in South Tyrol*
Anja Marcher, Ingrid Kofler, Elisa Innerhofer and Harald Pechlaner
*Tourism Geographies*, volume 24, issue 1 (2022), pp. 56–72

**Chapter 4**

*Local gastronomy, transnational labour: Farm-to-table tourism and migrant agricultural workers in Niagara-on-the-Lake, Canada*
Kristin Lozanski and Kayla Baumgartner
*Tourism Geographies*, volume 24, issue 1 (2022), pp. 73–95

**Chapter 5**

*Q method finds anti-refugee sentiments on Yemeni migration to Jeju*
Armigon Akhmedov, William Cannon Hunter and Jeong-Gil Choi
*Tourism Geographies*, volume 24, issue 1 (2022), pp. 96–116

**Chapter 6**

*"Traditional Mexican Midwifery" tourism excludes indigenous "others" and threatens sustainability*
Rosalynn A. Vega
*Tourism Geographies*, volume 24, issue 1 (2022), pp. 117–140

**Chapter 7**

*Exclave accessibility and cross-border travel: The pene-exclave of Ceuta, Spain*
Ioulia Poulaki, Andreas Papatheodorou, Alexandros Panagiotopoulos and Sotiroula Liasidou
*Tourism Geographies*, volume 24, issue 1 (2022), pp. 152–176

**Chapter 8**

*Labour migration and tourism mobilities: Time to bring sustainability into the debate*
Noel B. Salazar
*Tourism Geographies*, volume 24, issue 1 (2022), pp. 141–151

# Notes on Contributors

**Armigon Akhmedov** is Ph.D. Candidate at the Department of Hotel Management, Graduate School, Kyung Hee University, Seoul, Republic of Korea. His research interests include strategic management, corporate governance, and corporate social responsibility in the hospitality and tourism industry.

**Katsuhiro Ando** is Lecturer of the Faculty of Glocal Policy Management and Communication, Yamanashi Prefectural University, Japan. A Doctor of Engineering from Chiba University, his research interests are heritage tourism, community-based tourism and rural tourism in Vietnam and Japan. He worked as a tourism development consultant in Vietnam from 2003 to 2018, including Hoi An World Heritage Site and rural villages.

**Kayla Baumgartner** is PhD candidate in Geography at Western University. Her research interests include migration, gender, sexuality, and the social construction of space. She is particularly interested in how LGBT migrants "queer" the migration process itself and how they adapt to new environments. Kayla is currently finalizing her PhD dissertation, which looks at spaces of belonging for lesbian migrant women in urban South Africa.

**Huong T. Bui** is Associate Professor of the College of Asia Pacific Studies, Ritsumeikan Asia Pacific University (APU), Japan. She holds a PhD in Tourism Management from Griffith University in Australia. Her research interests are Asian youth tourism; Heritage conservation in Asia; Educational dark tourism in Japan; and Sustainable costal tourism in Vietnam.

**Jaeyeon Choe** is Lecturer at Glasgow School for Business and Society, Glasgow Caledonian University, UK and Visiting Professor at the School of Hospitality and Tourism at Hue University, Vietnam. Her primary research areas include the well-being of marginalised communities, social sustainability of destinations, and wellness/spiritual tourism in Southeast Asia.

**Jeong-Gil Choi** is Professor at the Department of Hotel Management, College of Hotel & Tourism Management, Kyung Hee University, Seoul, Republic of Korea. His research interests include finance, accounting, and forecasting in the hospitality and tourism industry.

**Ibán Díaz-Parra** studied Geography and Social Anthropology, completing his PhD in Human Geography at the University of Seville. He is currently Post-Doctoral Fellow at the University of Seville and secretary of the journal 'Habitat y Sociedad'. In the past, he was post-doctoral researcher at the National Autonomus University of Mexico and the

University of Buenos Aires as well as lecturer at the University of Cadiz. He is specialized in urban studies, conducting research on urban renewal and gentrification in Sevilla, Ciudad de Mexico and Buenos Aires.

**William Cannon Hunter** is Professor of Tourism in the College of Hotel and Tourism Management at Kyung Hee University, Seoul, Republic of Korea. His research interests include cultural tourism, smart tourism, semiotics and Q method. He is also highly involved in indigenous and climate change issues.

**Elisa Innerhofer** is Senior Researcher at the Center for Advanced Studies at Eurac Research in Bolzano. She studied international economics and business science as well as political science at the University of Innsbruck and the Marquette University in Milwaukee (WI, USA) and holds a PhD in Economics from the Catholic University of Eichstätt-Ingolstadt.

**Thomas E. Jones** is Associate Professor in Ritsumeikan Asia Pacific University (APU). His research interests include Sustainability; Protected Area management; and Nature-based tourism. After completing his PhD at the University of Tokyo, he worked for an NGO carrying out visitor surveys in the Japan Alps and monitoring Mount Fuji climbers during the summer season (2008–2017).

**Jaime Jover** is Post-Doctoral Researcher at the University of Seville, where he completed his PhD in Human Geography in 2017 and worked as research and teaching assistant between 2012 and 2016. He is interested in critical urban theory, social geographies and spatial planning, having studied cases across Andalusia and Portugal. He has previously held researcher and teaching appointments at the Centre for Geographical Studies of the University of Lisbon (Portugal) and Osuna College (Spain).

**Ingrid Kofler** is currently working as Senior Researcher at the Center for Advanced Studies at Eurac Research in Bolzano. Her research interests include innovation and networks, governance, migration and inter- and transdisciplinarity. She holds a PhD in Sociology from the University Paris Descartes-Sorbonne and University of Roma Tre.

**Sotiroula Liasidou** holds a PhD in Management (with emphasis on Tourism) from the University of Exeter, an MSc in Tourism Management from the University of Surrey and a BA (Hons) in Hospitality Management from Middlesex University. Currently, she is a Lecturer in the Department of Hotel and Tourism Management, Faculty of Management and Economics, at the Cyprus University of Technology (CUT).

**Kristin Lozanski** is Associate Professor of Sociology at King's University College at Western University in London, Canada with expertise in globalization, gender, and racialization. Her research interrogates migrant agricultural labour, transnational surrogacy, and birth tourism as disparate modes of transnational mobility to/from Canada, which overlap with reproduction, belonging, and citizenship.

**Peter Lugosi** is Professor of Culture and Organisation at the Oxford Brookes Business School and the Centre for Business, Society and Global Challenges at Oxford Brookes University. He has researched and published on a wide range of subjects including migrants' adaptation, labour market transition and entrepreneurship.

**Anja Marcher**, M.A., is former researcher at the Center for Advanced Studies at Eurac Research in Italy and is a doctoral candidate in Geography at the Catholic University of Eichstätt-Ingolstadt, Germany.

**Alexandros Panagiotopoulos is** a Transport Engineer, specialized in Railways and Choice Modeling. He has collaborated with EPFL, Lausanne as Scientific Assistant and with Swiss IMDM Infra as part-time consultant. Currently, he works in Aegean Airlines SA as an Analyst in the Customer Relations Management Department.

**Andreas Papatheodorou** is Professor in Industrial and Spatial Economics with Emphasis on Tourism at the University of the Aegean, Greece and Adjunct Professor at the University of New South Wales, Australia.

**Harald Pechlaner** holds a Chair in Tourism and is Director of the Center for Entrepreneurship at the Catholic University Eichstätt-Ingolstadt in Germany and is Adjunct Research Professor at Curtin University in Perth/WA (Curtin Business School). He is Head of the Center for Advanced Studies at the Eurac Research in Bolzano in Italy.

**Ioulia Poulaki** has been working in the Air Transport industry for 10 years with experience in commercial and operational departments. She also collaborates with Hellenic Open University, University of West Attica and University of Patras, as an Adjunct Lecturer in undergraduate and postgraduate programs of Tourism Management.

**Noel B. Salazar** is Research Professor in Anthropology at KU Leuven, Belgium. He is co-editor of *Tourism Imaginaries* (2014) and author of *Momentous Mobilities* (2018), *Envisioning Eden* (2010) and numerous peer-reviewed articles and book chapters on tourism and mobility. Salazar sits on the editorial boards of, among others, *Journal of Sustainable Tourism* and *International Journal of Tourism Anthropology*. In addition, he is past chair of the IUAES Commission on the Anthropology of Tourism and serves as an expert member for the European Parliament's Committee on Transport and Tourism (TRAN), the ICOMOS Cultural Tourism Committee (ICTC) and the UNITWIN-UNESCO Network 'Culture, Tourism and Development'.

**Rosalynn A. Vega** is Assistant Professor of Medical Anthropology and Global Health at the University of Texas Rio Grande Valley. She is the author of *No Alternative: Childbirth, Citizenship, and Indigenous Culture in Mexico* (2018).

# Introduction: Migration, tourism and social sustainability

Jaeyeon Choe and Peter Lugosi

**ABSTRACT**

In practice, the distinctions between tourism and migration are blurred. Tourism often drives various forms of mobility, and an international workforce is central to maintaining functioning tourism economies. This piece sketches out some critical themes and issues concerning intersections of tourism and migration, considering their relationships with and impacts on social sustainability. It highlights the contradictory ways in which tourism and migration are approached as political, social and economic phenomena. Whereas tourism is often viewed more positively, migration is recurrently politicised, and seen to challenge social systems and cultural values, despite the reliance of tourism on migrant labour. The discussion outlines the relevance of social sustainability to studies of migration and tourism. These include the need to assess how tourism planning, development and governance of tourism impacts on the sustainability of communities, which consequently influences attitudes towards migrants and tourists. It also reflects on how migrant-local connections may evolve, creating opportunities for positive, symbiotic co-existence, alongside exploitative relationships. It concludes by inviting further studies examining new forms and interactions between migration and tourism, which considers how research can contribute to greater social sustainability.

Migration and tourism have complex interactions, in some cases one driving the other, while in others, becoming competing practices and phenomena (Dwyer et al., 2014; Hall & Williams, 2002; Pappas & Papatheodorou, 2017; Williams & Hall, 2000). As global mobility intensifies, and tourism and migration become more deeply entangled, further research on their connections, complexities, critical issues and future directions is required. Arguably, our understanding of the intersections of migration and tourism is often obscured by western tourism concepts (Adams, 2020), with scholars and tourism authorities creating binary divisions between them. However, both arise out of a combination of social, economic and/or political factors, and mobilities are nurtured by tropes of imagination, with movement full of hopes, dreams, fears and uncertainties (Carling & Collins, 2018; Zhang & Su, 2020).

Migration and tourism do not emerge as consistent or predictable types of mobility, for example, with long-term migration from poor-to-rich countries and short-term tourism by the privileged elites. Scholars have increasingly recognised the disparate forms that international migratory processes can take, with transitory, incomplete, liquid, circular, seasonal and temporary manifestations (Collins, 2012; Hall & Williams, 2002; Hugo, 1982; Vosko et al., 2014). Long-term migrants, for instance, can become tourists as they travel back to their home countries for short trips, visiting friends and relatives, but also contemplate 'return' migration (Huang & Chen, 2020). There are also dual nationals, second-home owners, retirement migrants, lifestyle migrants, exchange students, contract workers, digital nomads, and working holiday makers who move between categories. 'Short term' is also a relative phrase in tourism, with some tourists becoming temporary or permanent lifestyle migrants (Benson & O'Reilly, 2009; Huete et al., 2013; Zaban, 2015). Therefore, migration or tourism both involves large-scale movement of people who engage with destinations in various ways and move between distinct categories.

As migration or tourism can be production or consumption oriented, mobility regimes are entangled with each other and other markets, such as retail, education, real estate, entrepreneurism, finance and healthcare. Tourism, like migration, is complex, with each destination having different spatialities, histories, actors, mechanisms, sites and spaces through which mobility is enacted, and where the possibilities for movement are linked to ethnicity, class, wealth and profession (Eisenschitz, 2016). While tourists and migrants intertwine, the global inequalities of bordering regimes and visa systems ensure that their mobilities and experiences also diverge. Most destinations, at least until recently, have risen to meet and support tourists from particular countries by providing them with visa free or low cost entry and broad protections. In contrast, migration has been problematised by divisive policy debates and emotive socio-political rhetoric. Consequently, migrants often have fewer rights than tourists in many destinations because of deep-seated inequalities (Abram et al., 2017; O'Regan, 2019).

There have been growing numbers of studies that link migration and tourism in contemporary societies (e.g., Baum, 2012; Bianchi, 2000; Janta et al., 2012; Lugosi & Allis, 2019; Ndiuini & Baum, 2021). These have examined topics such as the role of globalised capitalism and border management, migrant workers' well-being, social discrimination, inclusion, and settlement in host countries (Bianchi et al., 2020; Choe et al., 2020; Ladkin, 2011; Salazar, 2006). However, despite previous critiques (e.g. Williams & Hall, 2000), conceptualisations of the intersections between migration and tourism remain underdeveloped, with substantial gaps in exiting knowledge. Adams (2020), for example, argued that widely used terms and definitions of 'migration' and 'tourism' are primarily Western-centred, and more complex forms, practices and concepts should be explored within varying cultural and societal contexts. More work on long-term migration, temporary/seasonal migrant workers, lifestyle migrants, migrant workers in the gig economy, return migration and digital nomads is required.

A critical issue to explore within the migration and tourism nexus is how migrants, especially unskilled workers from developing countries, travel, work and live in destinations with limited legal protection, and worker rights. They often experience social discrimination, a lack of social support, inequalities and injustices (Bauder, 2012–13). Whilst many migrants relocate for work, family and/or lifestyle, many others have very little choice as to their spatial movements (Salazar, 2020). Bianchi et al. (2020) recommends adopting critical perspectives in exploring connections between migration and tourism to begin to tackle embedded structural inequalities and exploitation. Schiller and Salazar (2013, p. 189) point out that the current global order is beset by "different intersecting regimes of mobility" in which the movements of certain categories of people are "normalised" while others' movements are "criminalised". Events such as Brexit and terrorist attacks (such as in Paris in 2015), the emergence of wider popularist political movements, as well as the COVID-19 pandemic have accelerated criminalisation and immobilities (Bianchi et al., 2020; Salazar, 2020). Governments have therefore been required to implement innovative policies and practices to address the challenges of contemporary globalization to prevent fear and conflict (Melotti et al., 2018).

Migrant labour is often seen by governments and commercial operators as temporary and 'disposable,' which leads to precarious employment, social cleavages and economic inequalities (Robinson et al., 2019), as well as diminished contacts between migrants and locals. Moreover, media representations frequently perpetuate tensions by projecting negative images of certain minorities. Discriminatory profiling of national and ethnic communities further reinforces difference, resulting in greater social divisions (Bianchi et al., 2020).

However, migrant workers contribute significantly to economic development, cultural diversity, innovation, entrepreneurship, alongside knowledge transfer in tourism destinations (Lugosi & Allis, 2019; Williams & Baláž, 2014). Whilst tourism economies are largely measured through tourist arrivals and tourist receipts, the contribution of migrant workers is largely overlooked. Migration often generates tourist flows, through "the geographical extension of friendship and kinship networks" (Williams & Hall, 2000, p. 7). "Migration is an important determinant of VFR tourism and the relationship has progressively grown" (Dwyer et al., 2014, p. 15). Some destinations such as Dubai and Macao cannot sustain its tourism sector without migrant workers (Choe et al., 2020). As Macao's local labour is not large enough to fill positions in the tourism sector, they need to attract a high number of foreign migrant workers (Choe et al., 2020). Migrants also drive entrepreneurial activities utilising their cultural capital, for example, in the form of gastronomic knowledge, which often become part of the destination's culture (Farrer, 2021; Lugosi & Allis, 2019). Their contributions to a destination can sometimes become heritage attractions, and can help destinations display "cosmopolitan cultures, which are then used in place marketing and branding" (Janta & Lugosi, 2022, p. 3). Migration can thus transform a host society and impact on its cultural heritage and lifestyle. Migration flows have also contributed to the continuous reconstruction of national and destination identities and images (Melotti et al., 2018; Williams & Hall, 2000). However, while their contributions and impacts are sometimes valorised, the perceived value of migrants and migration is shaped by the changing character of many societies, which grapple with long-term challenges such as aging populations, restricted labour pools, alongside emotive political debates concerning nationality, identity and citizenship (Janta & Lugosi, 2022).

Research on labour and employment issues in countries undergoing rapid tourism economic development has expanded, but research has not caught up with tourism linked migration and social aspects of sustainability. Large numbers of migrant workers move to destinations such as Singapore, Hong Kong, and Dubai for employment; however, issues pertaining to social sustainability (e.g., well-being, integration, the distribution of power and resources, employment, education, the provision of basic infrastructure and services, freedom, justice, access to influential decision-making) have yet to be fully developed within tourism research.

Social sustainability in tourism research from 'host' and 'guest' perspectives is critical (Scheyvens, 2011; Zhang et al., 2017), and the nexus of migration and tourism is deeply connected to sustainability of host countries/destinations/communities. This concerns community cohesion, governance, place identity, as well as migrant workers' job security, legal protection, social welfare, well-being, and human rights. Social sustainability can be defined as "development (and/or growth) that is compatible with harmonious evolution of civil society, fostering an environment conducive to the compatible cohabitation of culturally and socially diverse groups while at the same time encouraging social integration, with improvements in the quality of life for all segments of the population" (Polese & Stren, 2000, p. 15–16). Often only seen primarily in economic terms, introducing the concept of social sustainability into the policy making of tourism destinations can improve locals' and migrants' lives and requires more attention in tourism studies.

This issue of *Tourism Geographies* aims to reflect upon, assess and develop our understanding of critical issues linking migration, tourism and social sustainability. Despite examining different socio-cultural contexts, the contributing authors strongly suggest that governments should innovate and implement policy changes to become strong and ethical leaders on migration and tourism, to positively influence social sustainability.

Building bridges between tourism and urban studies, in '*Who is the city for? Overtourism, lifestyle migration and social sustainability,*' Jover and Díaz-Parra examine the tensions between lifestyle migrants from wealthier countries, tourists and local residents in Seville as they share space and resources in the core historic district and urban centre. The authors argue that the rapid increase of tourism and the settlement of lifestyle migrants have resulted in social injustice and unsustainable practices for local communities as lower-income residents cannot afford to live in the area and locals have been gradually excluded from community spaces. While framing social sustainability as the capacity of indigenous communities to reproduce themselves, and preserving their social habits and customs, including avoiding physical displacement, the authors discuss how tourism potentially reshapes historic urban districts in socially unsustainable ways.

In the similar vein, but in a different part of the world – Vietnam – Jones, Bui, and Ando discuss the transformation, gentrification and social sustainability of Hoi An, a World Heritage Site (WHS), using the core-and-buffer-zone principle. In '*Zoning for world heritage sites: Dual dilemmas in development and demographics,*' the authors describe how massive international tourism inflows resulted in economic benefits including GDP growth, job creation, poverty alleviation and considerable funds for heritage conservation. However, the transformation has also shown negative socio-economic consequences such as price rises in real estate, rents and services for locals. The authors observe that erosion of local lifestyles is a consequence of a Euro-centric planning norm, and stress the need for a reappraisal of the conventional Eurocentric approach to zoning WHSs. By investigating the bipolarity between museumification of the core and concurrent development of the buffer zones, the study bridges a missing link between social sustainability and spatial planning. In doing so, the study contributes to our understanding of the social sustainability of a living heritage site.

In *'Perceptions of and interactions between locals, migrants, and tourists in South Tyrol,'* Marcher, Kofler, Innerhofer, and Pechlaner examine the tourism and migration nexus in South Tyrol, Italy, where a long history of tourism and multilingualism has attracted migrants. Their study found that the presence of migrants in a destination is strongly connected to the inflow of tourists while tourism further generates in-migration. Locals perceive tourists and migrants as 'non-local' or 'Others'. However, as migrants and tourists interact in increasingly complex ways with the local community, thus shaping it during their stay, locals tend to accept the 'new' population groups living, working and visiting the destination, and boundaries between them blur.

Lozanski and Baumgartner, in their piece *'Local gastronomy, transnational labour: Farm-to-table tourism and migrant agricultural workers in Niagara-on-the-Lake, Canada,'* explore the phenomenon of Mexican and Caribbean migrant workers travelling to Canada via a bilateral agreements to provide labour essential to the agricultural sector in Niagara-on-the-Lake (NOTL). This mobile workforce harvests local crops, which forms the basis of a significant gastrotourism sector. However, despite their key role, the structure of this transnational labour programme and the requisite aesthetics of tourism in NOTL render migrant workers' labour invisibility. This invisibility exacerbates the precarity of transnational workers within global capitalism, perpetuating wider societally unsustainable practices.

Beyond invisibility and precarity, Akhmedov, Hunter and Choi highlight the potential for more overt forms of discrimination and prejudice in their examination of public discourses and residents' attitudes towards migrants in South Korea. In *'Q method finds anti-refugee sentiments on Yemeni migration to Jeju,'* the authors suggest that interventions including structural shifts in policy or attitude changes through education are necessary to secure any form of social sustainability for resident-refugee relationships. The authors thus point to paths for finding practical solutions to 'host-guest' tensions.

In *'"Traditional Mexican Midwifery" tourism excludes indigenous "others" and threatens sustainability,'* Vega explores an expanding form of niche tourism to understand and challenge some of the underlying, intersectional issues threatening 'sustainability' in ethnomedical travel and consumption. She argues that neocolonial values and norms embedded in this type of tourism subtly reinforces romanticised stereotypes of the indigenous 'Other'. Ethnographic examples from multi-sited research in Mexico and Brazil demonstrate the usefulness of 'Traditional Mexican Midwifery' for critiquing existing misapplications of responsible tourism and proposing more sustainable futures.

Finally, in his commentary, *'Labour migration and tourism mobilities: Time to bring sustainability into the debate,'* Salazar stresses that migration and tourism are interconnected, similar but different forms of human mobility. Whilst tourism and migration often fuel each other, it is impossible to draw clear boundaries between the two because they constantly intersect, sometimes within one and the same individual. He notes that research on mobility often focuses merely on tourist movements in tourism studies while sustainability is directly related to the 'mobility' aspects of tourism-related labour. He adds that there has also been little detailed examination of the mechanisms that comprise and (re) produce the border-crossing movements of tourism labourers. If there is little attention to tourism-related labour mobilities, considerations related to the sustainability implications of migrant worker mobility are highlighted even less often. Salazar notes that despite the anthropocentric focus of sustainability as a concept, surprisingly little attention has been devoted to the aspect of social sustainability.

The contributions to this special issue provide thought-provoking insights concerning intersections of tourism and migration within the context of sustainable development. However, many gaps in existing knowledge remain (Santos et al., 2019). Tourism scholars are therefore prompted to further explore the impact of migrants and immigrant workers on destinations more widely, which extends to community relations, labour markets, place image and identities (Janta & Lugosi, 2022). Importantly, future research on migration and tourism, which considers the social sustainability of destinations, must remain conscious of the global unevenness and inequality in all forms of human movement and labour (Salazar, 2020). Scholars also need to understand migration and tourism in a more nuanced way, by examining the complex local situations, cultural dynamics and geopolitical issues (Adams, 2020). More broadly, there is work to be done to revise "Western-generated one-size-fits all tourism models, categories, and understandings" (Adams, 2020, p. 21). Finally, the COVID-19 pandemic crisis revealed and accelerated challenges around migration, tourism and social sustainability, thus stressing the need to develop impact-oriented research that extends across social, political and economic domains and utilises interdisciplinary approaches.

## Disclosure statement

No potential conflict of interest was reported by the authors.

## ORCID

Jaeyeon Choe http://orcid.org/0000-0003-3990-1730
Peter Lugosi http://orcid.org/0000-0002-7611-128X

## References

Abram, S., Feldman Bianco, B., Khosravi, S., Salazar, N. B., & de Genova, N. (2017). The free movement of people around the world would be Utopian. *Identities: Global Studies in Culture and Power*, *24*(2), 123–155. https://doi.org/10.1080/1070289X.2016.1142879

Adams, K. M. (2020). What western tourism concepts obscure: Intersections of migration and tourism in Indonesia. *Tourism Geographies*. https://doi.org/10.1080/14616688.2020.1765010

Bauder, H. (2012–13). Open borders: A utopia?*Justice Spatiale/Spatial Justice*, *5*, 1–10. Retrieved July 24, 2017, from http://www.jssj.org.

Baum, T. (2012). *Migrant workers in the international hotel industry*. Paper No. 112. International Labour Organisation. International Migration. Geneva.

Benson, M., & O'Reilly, K. (2009). Migration and the search for a better way of life: A critical exploration of lifestyle migration. *The Sociological Review*, *57*(4), 608–625. https://doi.org/10.1111/j.1467-954X.2009.01864.x

Bianchi, R. V. (2000). Migrant tourist-workers: Exploring the 'contact zones' of post-industrial tourism. *Current Issues in Tourism*, *3*(2), 107–137. https://doi.org/10.1080/13683500008667869

Bianchi, R. V., Stephenson, M., & Hannam, K. (2020). The contradictory politics of the right to travel: Mobilities, borders & tourism. *Mobilities*, *15*(2), 290–306. https://doi.org/10.1080/17450101.2020.1723251

Carling, J., & Collins, F. (2018). Aspiration, desire and drivers of migration. *Journal of Ethnic and Migration Studies*, *44*(6), 909–926. https://doi.org/10.1080/1369183X.2017.1384134

Choe, J., O'Regan, M., & Kimbu, A. N. (2020). Filipino migrant workers' leisure and subjective quality of life in Macao. *World Leisure Journal*, *62*(3), 242–260. https://doi.org/10.1080/16078055.2020.1798052

Collins, F. (2012). Transnational mobilities and urban spatialities: Notes from the Asia-Pacific. *Progress in Human Geography*, *36*(3), 316–335. https://doi.org/10.1177/0309132511423126

Dwyer, L., Seetaram, N., Forsyth, P., & King, B. (2014). Is the migration-tourism relationship only about VFR?*Annals of Tourism Research*, *46*(3), 130–143. https://doi.org/10.1016/j.annals.2014.03.009

Eizenberg, E., & Shilon, M. (2016). Pedagogy fort the new planner: Refining the qualitative tool box. *Environment and Planning B: Planning and Design*, *43*(6), 1118–1135. https://doi.org/10.1177/0265813515604477

Eizenberg, E., & Jabareen, Y. (2017). Social sustainability: A new conceptual framework. *Sustainability*, *9*(1), 68. https://doi.org/10.3390/su9010068

Enloe, C. (2014). *Bananas, Beaches and Bases: Making feminist sense of international politics*. University of California Press.

Farrer, J. (2021). From cooks to chefs: Skilled migrants in a globalising culinary field. *Journal of Ethnic and Migration Studies*, *47*(10), 2359–2375. https://doi.org/10.1080/1369183X.2020.1731990

Gillen, J., & Mostafanezhad, M. (2019). Geopolitical encounters of tourism: A conceptual approach. *Annals of Tourism Research*, *75*, 70–78. https://doi.org/10.1016/j.annals.2018.12.015

Hall, C. M., & Williams, A. M. (Eds.). (2002). *Tourism and migration: New relationships between production and consumption*. Springer Science & Business Media.

Huang, W. J., & Chen, C. C. (2020). Influence of transnational leisure on diaspora tourism among contemporary migrants. *Journal of Travel Research*. https://doi.org/10.1177/0047287520922315

Huete, R., Mantecón, A., & Estévez, J. (2013). Challenges in lifestyle migration research: Reflections and findings about the Spanish crisis. *Mobilities*, *8*(3), 331–348. https://doi.org/10.1080/17450101.2013.814236

Hugo, G. (1982). Circular migration in Indonesia. *Population and Development Review*, *8*(1), 59–84. https://doi.org/10.2307/1972690

Janta, H., Lugosi, P., Brown, L., & Ladkin, A. (2012). Migrant networks, language learning and tourism employment. *Tourism Management*, *33*(2), 431–439. https://doi.org/10.1016/j.tourman.2011.05.004

Janta, H., & Lugosi, P. (2022). Immigrant workers. In Buhalis, D. (Ed.) *Encyclopedia of tourism management and marketing*. Cheltenham.

Ladkin, A. (2011). Exploring Tourism Labor. *Annals of Tourism Research, 38*(3), 1135–1155. https://doi.org/10.1016/j.annals.2011.03.010

Lugosi, P., & Allis, T. (2019). Migrant entrepreneurship, value-creation practices and urban transformation in São Paulo, Brazil. *Revista Brasileira de Pesquisa em Turismo, 13*(1), 141–163. https://doi.org/10.7784/rbtur.v13i1.1538

Melotti, M., Ruspini, E., & Marra, E. (2018). Migration, tourism and peace: Lampedusa as a social laboratory. *Anatolia, 29*(2), 215–224. https://doi.org/10.1080/13032917.2017.1414441

Ndiuini, A., & Baum, T. (2021). Underemployment and lived experiences of migrant workers in the hotel industry: Policy and industry implications. *Journal of Policy Research in Tourism, Leisure and Events, 13*(1), 36–58. https://doi.org/10.1080/19407963.2020.1743487

O'Regan, M. (2019). Post-colonial Macau: Hope and despair in a world centre of tourism and leisure. *Via. Tourism Review.* https://doi.org/10.4000/viatourism.4469

Pappas, N., & Papatheodorou, A. (2017). Tourism and the refugee crisis in Greece: Perceptions and decision-making of accommodation providers. *Tourism Management, 63*, 31–41. https://doi.org/10.1016/j.tourman.2017.06.005

Polese, M., & Stren, R. (2000). *Social sustainability of cities: Diversity and the management of change.* University of Toronto Press.

Robinson, R. N., Martins, A., Solnet, D., & Baum, T. (2019). Sustaining precarity: Critically examining tourism and employment. *Journal of Sustainable Tourism, 27*(7), 1008–1025. https://doi.org/10.1080/09669582.2018.1538230

Rojek, C., & Urry, J. (1997). *Touring cultures. Transformations of travel and theory.* Routledge.

Salazar, N. B. (2006). ). Building a "Culture of Peace" through tourism: Reflex ive and analytical notes and queries. *Universitas Humanística, 62*, 319–333. Retrieved from http://www.scielo.org.co/pdf/unih/n62/n62a12.pdf

Salazar, N. B. (2020). Labour migration and tourism mobilities: Time to bring sustainability into the debate. *Tourism Geographies.* https://doi.org/10.1080/14616688.2020.1801827

Santos, R., Castanho, R. A., & Lousada, S. (2019). Return migration and tourism sustainability in Portugal: Extracting opportunities for sustainable common planning in Southern Europe. *Sustainability, 11*(22), 6468. https://doi.org/10.3390/su11226468

Scheyvens, R. (2011). The challenge of sustainable tourism development in the Maldives: Understanding the social and political dimensions of sustainability. *Asia Pacific Viewpoint, 52*(2), 148–164. https://doi.org/10.1111/j.1467-8373.2011.01447.x

Schiller, N. G., & Salazar, N. (2013). Regimes of mobility across the globe. *Journal of Ethnic and Migration Studies, 39*(2), 183–200. https://doi.org/10.1080/1369183X.2013.723253

Vosko, L. F., Preston, V., & Latham, R. (2014). *Liberating temporariness. Migration, work and citizenship in an era of insecurity.* McCill Queens University Press.

Williams, A., & Baláž, V. (2014). *International migration and knowledge.* Routledge.

Williams, A. M., & Hall, C. M. (2000). Tourism and migration: New relationships between production and consumption. *Tourism Geographies, 2*(1), 5–27. https://doi.org/10.1080/146166800363420

Zaban, H. (2015). Living in a bubble: Enclaves of transnational Jewish immigrants from West ern countries in Jerusalem. *Journal of International Migration and Integration, 16*(4), 1003–1021. https://doi.org/10.1007/s12134-014-0398-5

Zhang, H., & Su, X. (2020). Lifestyle migration and the (un) making of ideal home. *Geoforum, 115*, 111–119. https://doi.org/10.1016/j.geoforum.2020.07.001

Zhang, H. Q., Fan, D. X. F., Tse, T. S. M., & King, B. (2017). Creating a scale for assessing socially sustainable tourism. *Journal of Sustainable Tourism, 25*(1), 61–78. https://doi.org/10.1080/09669582.2016.1173044

# Who is the city for? Overtourism, lifestyle migration and social sustainability

Jaime Jover and Ibán Díaz-Parra

**ABSTRACT**

Seville, Spain's fourth largest city and the capital of Andalusia, is one of the most visited cities in the country and increasingly in Europe. Tourism has traditionally been an important sector for the urban economy, but after the 2008 financial crash it has become the city's main source of income. Tourist numbers have soared over the last decade and they are likely to keep growing now that the Lonely Planet has labelled Seville the best city to visit in 2018 and Airbnb has chosen Andalusia as one of the top destinations for 2019. There has also been an increase in the number of migrants from wealthy countries over the last few years. Not surprisingly, these migrants and tourists share spatial patterns: both generally concentrate on Seville's historic district and urban centre. We argue that the rapid increase of tourism and the settlement of lifestyle migrants in this part of the city have triggered spatial processes that have resulted in social injustice. Therefore, we explore whether the growth of visitor numbers has reached the extent of overtourism, because the tourist infrastructure and its activities may clash with those of local residents. We also study whether the socio-economic profile of many of these residents has changed due to migration, which could potentially mean the beginning of a transnational gentrification process. Seville's city centre is becoming a privileged place, where the average Sevillian cannot afford to live and increasingly feels out of place. Our argument is supported by a combination of quantitative data analysis (demographic, housing and tourism statistics) and qualitative research (interviews) analyses. Our case illustrates how tourism helps to reshape historic urban districts and becomes socially unsustainable, as it gradually excludes local communities.

摘要

塞维利亚是西班牙第四大城市，也是安达卢西亚的首都，是西班牙最受欢迎的旅游城市之一，在欧洲也越来越受欢迎。旅游业传统上一直是该市重要的经济部门，但在2008年金融危机之后，它已成为该城市的主要收入来源。在过去的十年里，游客数量激增，现在可能还会继续增长，因为《孤独星球》将塞维利亚列为2018年最佳旅游城市，而Airbnb将安达卢西亚列为2019年最佳旅游目的地之一。过去几年来，来自富裕国家的移民数量也有所增加。毫不奇怪，这些移民和游客有共同的空间模式:通常都集中在塞维利亚的历史街区和城市中心。我们认为，旅游业的快速增长和该街区生活方式移民的定居引发了导致社会不公的空间进程。因此，我们探讨游客数量的增长是否已经达到了过度旅游的程度, 因为旅游基础设施及其活动可能与当地居民生活设施与及日常活动有冲突。我们还研究了这些居民的社会经济状况是否因移民而改变, 这可能意味着跨国绅士化进程的开始。塞维利亚的市中心正在成为一个特权场所, 在那里, 普通的塞维利亚人无法生活, 并越来越感到格格不入。我们的论点得到了定量数据分析(人口、住房和旅游统计)和定性研究(访谈)分析的共同支持。我们的案例说明, 旅游业如何帮助重塑历史城区, 并在社会上变得不可持续, 因为它逐渐排斥当地社区。

Sevilla, la cuarta ciudad de España y capital de Andalucía, es uno de los destinos más visitados del país y cada vez más de Europa. El turismo ha sido siempre un sector clave en la economía urbana, pero tras la crisis de 2008 se ha convertido en la principal fuente de ingresos. El incremento del turismo en la última década no tiene precedentes, y probablemente siga aumentando tras las designaciones de Lonely Planet como la mejor ciudad para visitar en 2018, y la de Airbnb, que ha elegido a Andalucía como uno de los destinos más apetecibles en 2019. Al mismo tiempo, los migrantes de países ricos han crecido en los últimos años. Estos comparten con los turistas un patrón espacial claro: ambos se concentran en el centro urbano e histórico de Sevilla. Se parte de que el rápido aumento del turismo y la llegada de migrantes por estilo de vida en esa parte de la ciudad ha disparado procesos espaciales, profundizando en injusticias sociales. Por tanto, investigamos si el crecimiento de visitantes ha alcanzado un punto de *sobreturismo* en tanto que las infraestructuras y actividades turísticas compiten con aquellas de los residentes. También estudiamos si el perfil socio-económico de muchos de estos residentes ha cambiado debido a la migración, lo que podría estar provocando un proceso de gentrificación transnacional. El centro de Sevilla se está convirtiendo en un lugar privilegiado donde el sevillano medio no se puede permitir vivir y con mayor frecuencia se siente incómodo. Nuestra investigación se basa en una combinación de análisis cuantitativos (a través de estadísticas sobre demografía, turismo y vivienda) y cualitativos (mediante entrevistas). El caso evidencia cómo el turismo ayuda a reconfigurar los distritos históricos turísticos y puede convertirse en socialmente insostenible en tanto que progresivamente excluye a las comunidades locales.

## 1. Introduction

'Tourism kills the city' or 'tourists go home' have become commonplace stickers and graffiti in southern European urban destinations, prompting debates about 'tourismophobia' (Huete & Mantecón, 2018; Milano, 2018). Such fears aside, it is undeniable that social conflict has escalated as tourism has rapidly increased across Mediterranean cities and tourist arrival numbers have broken records. Protests and mobilisation have increased in these destinations, with the emergence of urban social movements that denounce tourism's negative effects or touristification, i.e., tourism-related housing and real estate speculation, congestion and the privatisation of public spaces, the appropriation and commodification of culture or environmental issues, among others (Jover et al., 2018). A network of European territories – mainly cities –

coping with problems in highly popular tourist destinations was established in 2017: the SET (Southern Europe facing Touristification) network. In this paper, we focus on two interrelated processes: overtourism and lifestyle migration, and we examine how they are related to the displacement of residents from historic city centres.

The impact of tourism on coastal areas has been widely studied in the past, especially in 'sun and sand' destinations (Blázquez & Murray, 2010), while studies about the mobility and long-term settlement of expats in search of a better quality of life have also increased, leading to the conceptualisation of lifestyle migration (Benson, 2012; Hayes, 2018). Research has also focused on urban tourism destinations (Ashworth & Page, 2011), among them cultural destinations, ranging from large metropolises to medium-sized cities with a tourist appeal based on heritage and important historic city centres (Calle Vaquero, 2006; Richards, 1996). In this sense, Venice has become a paradigmatic case, where the massive influx of visitors has led to overtourism (Seraphin et al., 2018). This situation, we think, may be affecting the social sustainability of local communities by displacing lower-income families, saturating public spaces or decreasing the quality of labour, the latter being a primary concern in contexts where employment is highly seasonal, wages are low, workers are not usually unionised and can potentially suffer from marginalisation (Cañada, 2018; Cheer, 2018). Tourism is receiving growing attention from political economy perspectives (Bianchi, 2018), where critical urban scholars stand out, focusing on the relationship between tourism and urban planning, local politics or socio-spatial processes such as gentrification (Cócola-Gant, 2018; Sequera & Nofre, 2018). Although these topics have already been discussed in cities with an economy that relies on the tourist sector, the shift towards a sort of tourism dependency after the 2008 financial crisis and the austerity regime imposed in the European Mediterranean context require greater attention (Annunziata & Lees, 2016). We believe research must focus on broadening the cases of study, as well as on building bridges between tourism, migration and urban studies, putting them into dialogue and learning from one another. In this regard, we wonder what the relation of tourism is with classic and transnational gentrification? Are overtourism and lifestyle migration leading to the effective displacement of residents in increasingly touristy urban areas? And are they threating the possibilities of social reproduction of local communities?

We answer these questions by establishing comparisons between the increasing flow of visitors and the tourist infrastructure, the arrival of lifestyle migrants and the displacement of local residents in a destination that specialises in cultural tourism. We focus on Seville, Spain's fourth largest city and the capital of Andalusia, the most populated yet most underdeveloped region in the country, recently labelled one of Airbnb's top destinations for 2019. Seville has received record breaking numbers of visitors in recent years, with the latest record being more than 3 million visitors in 2018 thanks to marketing campaigns that managed to secure Lonely Planet's endorsement of Seville as the world's best city to visit that year, among other achievements. Unlike Madrid or Barcelona, Seville has not received as much scholarly attention, despite being one of Spain's main tourist destinations. Our hypothesis is that current trends in tourism-led urban economies are making traditional residential neighbourhoods within Seville's historic district socially unsustainable. In the Mediterranean context, historic urban areas suffered disinvestment and population loss during the 1960s, 1970s and 1980s. Before the end of the twentieth century there was a reversal in this trend as a result of urban renewal policies that usually translated into gentrification. We suggest that tourist specialisation jeopardises local communities because it undermines the possibilities of reproduction of long-standing families who are subject to displacement at an even quicker pace than that caused by previous waves of local middle and upper class gentrifiers. Therefore, we explore how different processes taking place are overlapping in a specific urban area over the past twenty years.

## 2. Background: overtourism and lifestyle migration as socially unsustainable in urban cultural destinations

Social sustainability is a highly dynamic concept that brings together economic and social aspects, as well as environmental issues. Polese & Stren highlight that cities must be socially sustainable if they are to be environmentally sustainable, defining the former as the adaptability between economic growth and the 'harmonious evolution of civil society, fostering an environment conducive to the compatible cohabitation of culturally and socially diverse groups while at the same time encouraging social integration' (2000, pp. 15–16). Dempsey et al. (2009) define urban social sustainability through two key concepts: the sustainability of community and social equity, with the latter referring to a society 'in which there are no "exclusionary" or discriminatory practices hindering individuals from participating economically, socially and politically in society'. Therefore, social equity also relates to social justice, while the sustainability of the community points to its capability for social reproduction. Taken together, we frame social sustainability as the capacity of local communities to reproduce themselves, preserving their social habits and customs, which includes avoiding physical displacement. Such displacement is deeply unjust as locals are uprooted from the place where they live and have personally developed in many cases for years or decades.

Here there is a connection with gentrification as a process that necessarily entails the displacement of local groups from their neighbourhoods. Gentrification involves population substitution by residents with a higher income and normally a higher status as well and is essentially explained by the rule of the market in capitalist cities, which comprises the logics of capital accumulation and its facilitation by public institutions under neoliberalism (Smith, 1996, 2002). When these newcomers are foreign middle and upper classes, we encounter a process of transnational gentrification. This has been addressed in urban contexts such as historic districts in Latin America, where it is driven by lifestyle migrants from the US and Canada (Hayes, 2018; Sigler & Wachsmuth, 2015). Aside from their higher purchasing power and the better quality of life that this permits, these places attract them in a similar way as they attract visitors: by offering a unique and authentic urban culture, commonly stemming from material and/or immaterial heritage. In a globalised world, where standardisation of production and culture homogenisation are rapidly increasing at all scales, differentiation has become fundamental. This argument sustained Zukin's (1995) idea of culture as 'the business of cities', because culture cannot relocate whereas it can be easily commoditised and, as an expression of identity, it helps reinforce the social order. Manufacturing symbols, images as well as places – especially in cultural tourism destinations – are key to understanding contemporary urban processes that ultimately contribute to the production of inequalities across the cityscape.

In addition to transnational gentrification, scholars studying these phenomena have coined the term tourism gentrification. In this sense, neighbourhood upscaling and socio-spatial segregation in urban areas popular with tourists is a deviation from gentrification (Gotham, 2005, Cócola-Gant, 2018). Although we understand that current waves of urban tourism are strongly interrelated with gentrification theory as regards market logics (such as the Airbnb rent gap, see Wachsmuth & Weisler, 2018) and neoliberal spatial strategies, and as much as tourism could be a key driver in neighbourhood upscaling, their effects differ slightly. The intensification of tourist activity in a specific area cannot be strictly identified as gentrification because there is not population replacement. Rather, touristification seems to be a more accurate term to define the process of rapid change in tourist-led urban economies and its consequences, also taking into consideration issues such as retail transformation, public space appropriation or the quality of air in relation to the enlargement of tourist infrastructures such as airports and cruise quays. Actually, the latter has become one of the sticking points of the aforementioned social movement network against touristification, which has recently signed the Stay Grounded declaration against air traffic expansion (Assemblea de Barris per un Turisme Sostenible, 2019). However, gentrification and touristification share the most important consequence: the displacement of lower-income households.

Overtourism is another way to refer to tourism intensification within and around urban areas. The World Tourism Organization defines it as 'the impact of tourism on a destination, or parts thereof, that excessively influences perceived quality of life of citizens and/or quality of visitor experiences in a negative way' (UNWTO, 2018a, p. 4). Likewise, Goodwin (2017, p. 1) says that overtourism 'describes destinations where hosts or guests, locals or visitors, feel that there are too many visitors and that the quality of life in the area or the quality of the experience has deteriorated unacceptably'. Both definitions focus on perception, which is key to understanding the conceptualisation of overtourism, without forgetting that quantitative data on tourism (travellers, revenues, expenses, etc.) has not stopped growing worldwide over the past few years (see UNWTO yearly data reports). If touristification refers to the consequences of that increase in urban contexts, overtourism deals with the resulting scenario of the process, where two ideas must be highlighted. Firstly, it is not just the number of visitors, but the space they use and how they are distributed. It goes beyond traditional conceptions of tourism congestion or saturation; overtourism refers to a more severe situation. Tourism concentrates in specific urban areas where public policies, from city marketing to heritage restoration, and private investment converge. These urban hotspots accommodate most of the city's tourist infrastructure, both demand and offer, including public transportation and facilities, tourist amenities and heritage sites, hotels and hostels, short-term holiday rentals, restaurants, cafés and pubs as well as any other tourism-oriented leisure businesses. A spatially uneven distribution leads to touristification in an area that generally is or at least used to be mainly residential, so urban functions come into conflict, potentially spreading to surrounding neighbourhoods too. Secondly, overtourism occurs in a context where tourism has become monopolistic, i.e., the majority of new businesses that open are related directly or indirectly to tourism, in line with public policies that focus on attracting capital and visitors. The high specialisation also means the restructuring of the urban economy, potentially dismantling other sectors, which could go beyond the city limits and spread across its hinterland (Murray, 2015). In this context locals' perceptions of tourism change, following previous definitions. However, far from being a uniform phenomenon, overtourism develops differently depending on the local, regional and national contexts.

We agree with both the UNWTO (2018a) and recent research (Koens et al., 2018) that has pointed out that overtourism is not a tourism-only problem. The argument is that tourism competes with residential functions or urban mobility to the extent of making them difficult or blocking them. Unlike places such as Disneyland that were designed for tourism and leisure purposes, cities were not initially built with these functions in mind. Their urban value and signification as heritage and therefore their transformation into cultural assets and participation in the accumulation of capital through tourism came afterwards (Eramo, 2014). Historic urban areas have joined tourist circuits, bringing about pressure on their residential function and social infrastructures, and jeopardising local communities. Better-off foreigners who profit from the atmosphere created for tourist purposes and the related facilities further intensify this process, despite there being differences between migrants and visitors in terms of adjusting to the urban culture, understanding the city and behaving accordingly. We must distinguish between those foreigners who spend a few hours or nights in a city and those who stay longer, because their integration, i.e., their relationship with the space and their own social bonds, is not the same. However, there will of course always be particularities amongst a wide range of cases: there will be lifestyle migrants who do not wish to integrate or tourists who are highly sensitive towards the places they visit.

Overtourism remains under-theorised in academic literature. As Milano (2018) and Koens et al. (2018) posit, tourism's negative effects are not new; rather it seems as if old concerns have been reshaped. Nevertheless, this idea is disputable after the restructuration of capitalist cities towards expanding the tourist sector in the context of crisis, especially in the Mediterranean. The proliferation of digital platforms such as Airbnb, Booking.com or HomeAway, the expansion of low-cost airlines or the deseasonalisation of tourist destinations are factors that explain this restructuring. At the same time, the response to the spread of tourism in urban areas has reached a new social climax, as we have already mentioned. Solutions must be innovative, and so far have focused on regulating the short-term holiday rental market, which has been identified as one of the main issues causing social unrest (Blanco-Romero et al., 2018). Tourism's growing social unsustainability needs examination given its unfair consequences to local populations and the way in which it combines with processes driven by privileged local groups, which could even threaten those communities created after previous processes of gentrification. Moreover, the damage these phenomena inflict on the destination is worth studying, as it has the potential for turning into intolerance and anger towards tourism, and therefore not just destroying part of the city's social and cultural fabric, but also its condition as an economic resource. This is why we address the relationship between overtourism and lifestyle migration linked with processes of gentrification and touristification through the lens of social sustainability, in order to contribute to debates on tourism, migration, and urban studies.

## 3. Methodology and sources

We combine quantitative and qualitative methodologies at two scales of analysis: the national and regional, in particular the city. The first analysis serves as contextualisation for the case study of Seville and draws on tourism, migration and housing statistics (from the 2001 census to the most recent data available), coupled with the main bibliography on the topic. At the city scale, we begin by giving a general overview of Seville's historic district from the beginning of the twenty-first century, using academic literature and planning documents. We have also identified and assessed city sights by analysing local guides and the promotion they receive on Seville's official tourism website (Turismo de Sevilla: https://www.visitasevilla.es/turismo-de-sevilla). We then move on to examine overtourism and lifestyle migration as being potentially socially unsustainable for Seville's historic district. As part of this, we draw on a different set of data to compare the evolution of Seville as a whole in relation to the historic district. We explore demographic changes through variations in total population, age and level of studies, which could entail demographic revitalization or decline in relation to gentrification. Lifestyle migration numbers come from data on residents from wealthy foreign countries. Data on population is readily available from the yearly city statistics, whereas data on residents' nationality or level of studies are restricted to census years (2001 and 2011). We also examine overtourism through statistics that indicate the specialisation in the area studied, focusing on the total increase in tourists and the evolution and distribution (applying a Kernel method) of the total number of tourist accommodation establishments and beds. To do so, we use the Andalusian Tourism Register and the collaborative DataHippo project (https://datahippo.org/en/), which gathers data directly from Airbnb and HomeAway websites.

Quantitative analyses overlap with responses from 17 in-depth interviews we carried out with people from different backgrounds, divided into three profiles. Firstly, we conducted 6 extensive interviews with current foreign residents of the area of study (EA), who have been living there for between 2 to 38 years, and who originally come from France, Germany, Italy and the Netherlands. They all fit the description of a lifestyle migrant, not just because they come from a wealthier country, but also because they decided to settle in Seville due to the cultural appeal of the city and especially of its historic district. Secondly, we interviewed 6 former residents of Seville's historic district, all aged under 50 who have moved or have been displaced in recent years (EB). All share a similar socio-demographic profile: adults living in smaller households with highly skilled members, as became predominant in the district at the beginning of the twenty-first century. They have another common factor: all still live in Seville. When they decided or had to move out of the district, they looked for a location in nearby neighbourhoods so they would be as close as possible to where they used to reside. Because of this, we can reject the idea that their cases fit the cycle of creation of new households in the transition from young adults to midlife, which usually involves movement from the inner city to suburban locations and entails a completely different process. Within this group of former residents there are two Sevillians, two Spanish non-Andalusians (originally from Madrid) and two foreigners (Argentinian and French). Thirdly, we interviewed 5 Spanish residents (EC) who moved into the historic district during the period of study. These two last groups of participants helped us to understand the historic district's demography (revitalisation or decline) and its connection to gentrification. All three profiles were also asked about how they have experienced the transformation of the historic district as a tourist destination, conflicts between residential and tourist functions, as well as factors of displacement. The recruitment strategy for finding our interviewees followed the snowball effect: the researchers talked to or sent emails to friends and acquaintances, asking for contacts in their networks that would fit any of the three categories. Out of all the responses, the researchers selected those who met the requirements, aiming for background diversity (in terms of age, nationality, level of studies, years residing in Seville) within each of the profiles.

## 4. Contexts

### *4.1. National: Spain and Andalusia as a destination for foreigners*

Tourism is undeniably one of the most important economic sectors worldwide, gaining weight yearly. In global terms, Southern Europe increased its visitors' figures by 12.8% in 2017, partially sustained by Spain's growth in international tourism arrivals, with 8.6% more visitors than the previous year. A total of 81.8 million tourists visited Spain in 2017, the second most popular destination in the world (UNWTO, 2018b). However, there is an uneven distribution of tourism within the country and some regions such as Andalusia have a higher specialisation than others. Reasons are found in recent history. Spain began specialising in tourism during the rule of the fascist dictator Francisco Franco (1939–1975), particularly from the 1960s onwards when 'desarrollismo' began. As Murray (2015) has studied, Spain became a perfect target for recycling international capital by investing in real estate and at the same time satisfying an increasing tourist demand. Spain's transition towards a liberal democracy in the early 1980s coincided with the rise of neoliberalism, and therefore the economy restructured sharply towards services, deepening its specialisation in tourism and real estate. The development of both sectors is key to understanding Spain's and Andalusia's most recent periods of economic growth and crisis, which illustrates the fluctuation of foreign investment in real estate (Jover & Díaz-Parra, 2019) or hotel financing (Yrigoy, 2016).

Following this growth there has been a greater presence of foreigners in both Spain and Andalusia. At the end of the twentieth century, migrants did not even make up 3% of the Spanish population, while in 2011 they represented almost 12%, with a similar figure being registered for Andalusia (data taken from Population and Housing Censuses, Spanish Statistics Office). As the twentieth century drew to a close, most migrants arriving in Spain came from African and Latin American countries, whereas the first years of the new century witnessed a rise in the number of migrants from EU states. As a result, there has been an increase in foreigners purchasing houses in Spain, depending on the economic situation. In 2006, almost 9% of the market share in real estate belonged to migrants, with British buyers being the largest group. Sales to Ecuadorians, Moroccans and Romanians also increased, following the growth of these nationals settling in the country (data taken from College of Spanish Property Registrars statistical yearbooks). In 2010, as the real estate bubble burst, housing prices continued to be relatively high and unemployment grew, the Spanish migratory balance reversed and foreign housing buyers fell to 4.45%, with British, French and German buyers still being active. New buyers have emerged from Russia and Nordic countries since 2012, with a particularity: the international share of the market grew, while the numbers of total dwellings sold fell by more than 100,000 compared to 2010. Although house purchases have not recovered their pre-crisis peak, since 2016 the market has been reactivated, in part due to migrants who now represent 13.11% of the total number of buyers (13.84% in Andalusia). These foreign purchasers, mainly from wealthy northern European countries, have been fundamental for the Spanish real estate market during the crisis years. Among them there are many investors, but also lifestyle migrants who buy a second home, especially in regions such as Andalusia.

### *4.2. Local context: Seville's historic district as a tourist space*

Seville's tourist success is not entirely novel. In 1929 the city hosted the Ibero-American Exhibition with the purpose of restoring its commercial ties with Latin America. As a result, urban restructuring was intense, significantly in the Santa Cruz quarter (Díaz-Parra, 2016). The latter became one of the main tourist attractions after the slight opening of Franco's regime in the 1960s. By that time, Seville's urban growth had surpassed the city walls and the historic district's population declined. That trend continued for decades as the city spread, but distribution was socio-spatially uneven (Cruz, 1986). In general terms, the historic district's northern half was traditionally where housing for the working classes was located, as well as low intensity industries, which were socially stigmatised and dismantled respectively through the second half of the twentieth century. Despite strong depopulation, density was still higher than in the southern half, where the council fostered tertiary uses. This area became the city's shopping precinct, with its public spaces being turned into commercial streets that shared this function with administrative or religious buildings, which had been located in this southern part for centuries. It has traditionally been recognised as the city centre (Villar Lama & Fernández-Tabales, 2017). The main tourist attractions are located here, including the city's UNESCO World Heritage sites: the Cathedral, the Royal Palace and the Archives of the Indies.

The listing of these monuments dates back to 1987 when the first City Master Plan under democracy was approved. This document guaranteed the largest urban restructuring since 1929, coinciding with another world event: the Universal Expo in 1992. After decades of neglect, the exhibition was an excuse to recover the city's cultural heritage and boost tourism. Therefore, urban planning concentrated its

**Figure 1.** Distribution and intensity of tourist attractions in Seville.
Source: Authors based on the official Seville tourism website and fieldwork.

efforts on renovating those quarters that had suffered from abandonment in the historic district's northern area, leading to strong speculative tensions during the 1990s and resulting in gentrification dominated by local middle classes (Díaz-Parra, 2010, 2014). The subsequent City Master Plan of 2006, which is still in effect today, along with city-brand strategies (Seville City Council, 2003, 2006) elaborate on this idea of turning the whole historic district and more especially its northern half into a privileged, tourist-friendly space (Jover & Díaz-Parra, 2019; Jover, 2019). Among others, public policies have allowed new tourist infrastructures, reduced to a minimum the few existing public housing developments or adopted a flexible idea of restoration when it comes to renovating historic buildings. Nonetheless most of the tourist infrastructure can still be found in the southern half. Figure 1 shows the importance of this area regarding the distribution of the main tourist attraction hubs on three levels of intensity.

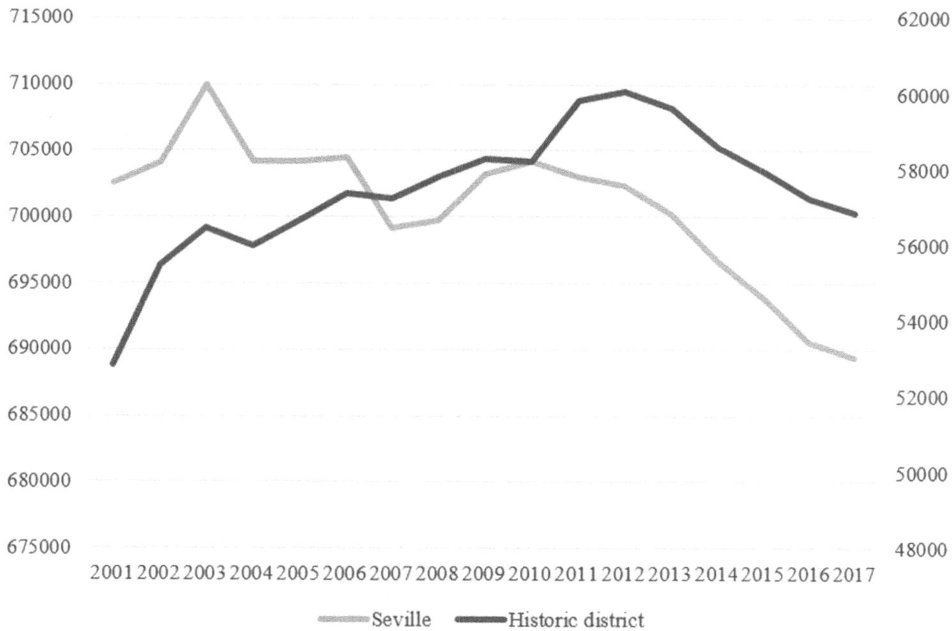


**Figure 2.** Total population in Seville and its historic district (2001–2017).
Source: Authors from Seville City Council Statistics Service.

## 5. Findings: changing dynamics within Seville's historic district

### *5.1. Revitalization and gentrification*

The trends within Seville's historic district of population decline and aging that had started in the 1990s reversed in the new millennium. Figures 2 and 3 compare changes in total population and the elderly rate respectively between Seville and its historic district since the beginning of the twenty-first century. At this time the population grew by more than 13%, while the elderly rate (the segment of population older than 65) decreased by nearly 3%. Rejuvenation was the result of new younger households, because elders' absolute numbers remain stagnated. This is especially meaningful once Seville as a whole had the opposite tendency: stationary population growth and the elderly rate rising by 2%. Therefore, Seville city centre's particular dynamic responded to the settlement of new residents with a specific profile: young adults with high levels of education and highly skilled jobs. According to the national censuses, numbers of university graduates in the district between 2001 and 2011 grew from 19.79% to 43.02%, while the illiteracy rate fell from 15.9% to a marginal 2.68% (Spanish Statistics Office). But, as the economy started to recover after the 2008 financial crisis, the tendency changed. In 2012, the historic district began losing population yearly, while the elderly rate grew, ranking almost equally with that of Seville as a whole. Again, this rise does not go hand in hand with an increase in the number of elders, so it can only mean the displacement of young households.

We have identified a first wave of local middle-class gentrification approximately between 1995 and 2012, which is confirmed by current and former residents who we interviewed and who settled in the district during that period. As regards their

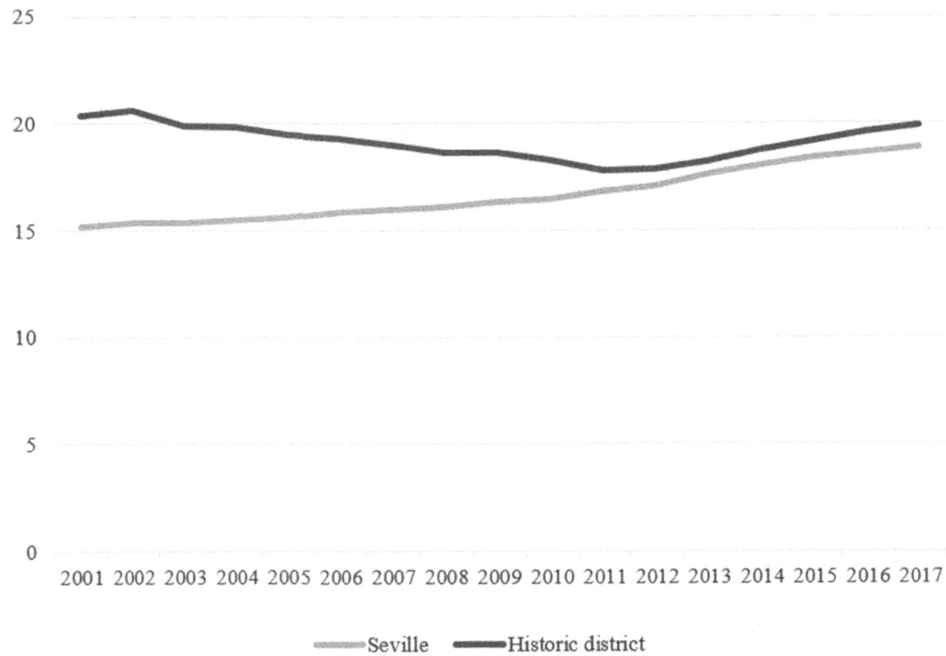


**Figure 3.** Elderly rate in Seville and its historic district (2001–2017).
Source: Authors from Seville City Council Statistics Service.

reasons, they generally highlight the existence of a community vibe that either existed previously or was created by the new settlers. Nearly all the Spanish respondents mentioned the pre-existence of a network of friends living there, as well as shops, pubs, squares and other spaces of socialisation. Another topic in line with the community feel is the conception of the area as a 'barrio', underlining its own personality and the close and friendly relations between residents. Participants refer to people frequently greeting each other in the streets, shops and pubs where they would go any time with a high probability of bumping into friends. In their opinion, cultural life (bookshops, cafés with live music, cinemas, etc.) also gives personality to the historic district and is a factor in attracting new residents. EB4 said: 'I liked the aesthetics too. I came from Madrid, huge and impersonal, and I enjoyed living in a "barrio" of little cute houses… I had the feeling of living in a story, a special place, authentic, as if I were on a permanent vacation'.

Besides the socio-cultural element, other reasons to live in the area are proximity to their jobs, mostly highly skilled (EB1 works in an architecture studio, EB3 and EB4 are studying at the university) or owning a pub (EB2 and EC2). Every interviewee who settled before 2010 made it clear that the area offered the opportunity to stay. They usually refer to large and roomy flats with relatively low rents, and some of them who are owners after buying in the 1990s (EC1 and EC4) also remember reasonable housing prices. EC1 is a civil servant who moved in at the end of the 1980s, when:

> We used to call the neighbourhood [San Luis, in the district's northern area] Beirut … they developed new buildings, renovated old ones, streets, squares… I imagine they expelled long-standing residents or the older ones progressively died, and then more

expensive flats began to appear ... We were lucky because we bought in 1988. My brother bought an adjacent flat, very cheaply. He sold it a few years later, making a high profit, because he wanted a larger one to have a family. Already in the 1990s there was a boom on flat prices

EC1 also makes reference to the social strata that used to live in her neighbourhood: 'Before there were a lot of low-class people, drug users... There were many *corrales de vecinos* [old collective tenants' houses that were very common in Andalusian cities]. Some were lifelong residents and others were homeless or squatters. I saw the police foreclosing entire buildings'. EB1 depicts and summarises a similar idea. She moved in the early 2000s and left in 2016 and suggested two processes of population substitution: 'Initially, only people with low salaries or even marginal people lived here ... they were displaced, sometimes in a violent way. We, younger people, came because renting was cheap. I think we tried to live together with old residents. The second substitution process is of residents by tourists'.

### *5.2. Overtourism*

The growing tourist appeal of Seville's historic district is a recurrent theme in almost all interviews. A recurring idea is that, as the gentrification frontier – using Smith's (1996) conceptualisation – has expanded, so has tourism, which has not stopped growing in Seville since the beginning of the century. In 2001, there were approximately 450 tourist accommodation establishments: 150 hotels and hostels and 300 holiday apartments. The latter are usually inexpensive short-stay lodgings that abide by the tourist legislation and include a series of rooms within the same vicinity,

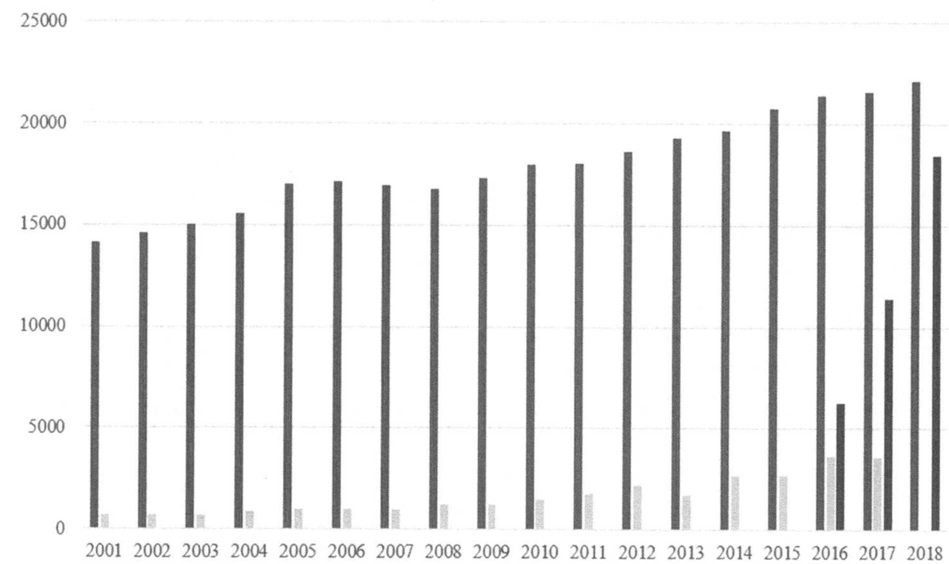


**Figure 4.** Tourist beds in Seville (2001–2018).
Source: Authors from Spanish Statistics Office (for hotels & hostels) and Andalusian Tourism Register (for tourist apartments and homes).

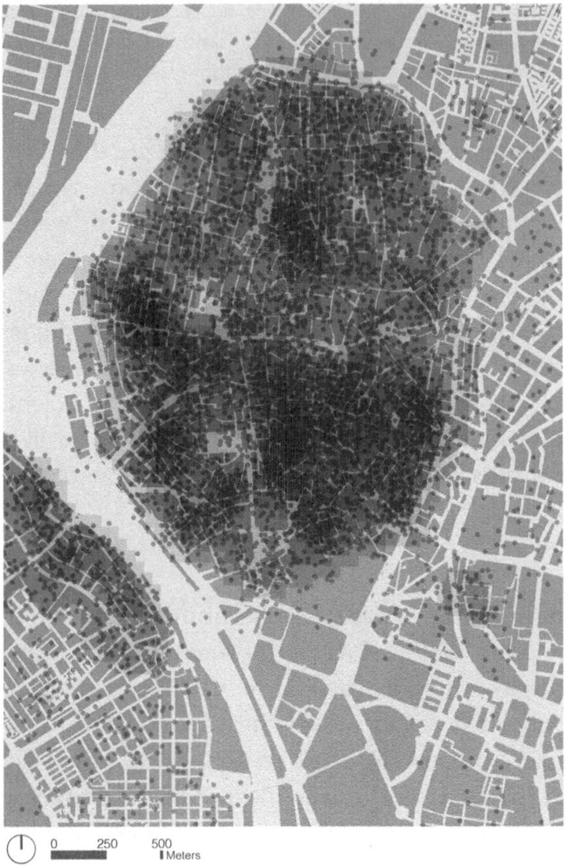


**Figure 5.** Distribution and density of (a) tourist homes and (b) hotels in Seville historic district in 2018.
Source: Authors from DataHippo (for tourist homes) and Seville City Council spatial data infrastructure (for hotels).

normally an entire building – guesthouses would be a good example. Hotels and hostels had the largest share of estimated tourist beds in 2001: 14,131 compared to just 700 in the case of apartments. Figure 4 shows official data on the evolution of tourist beds over the past few years. Seville's accommodation offer stagnated between 2006 and 2009, coinciding with the worst years of the crisis, when national arrival numbers dropped dramatically. However, since 2010 there has been an increase in the number of hotel and hostel beds again. Apartments have behaved differently: there was a slight decrease in 2013, precisely a few months after Airbnb landed in the Spanish market. A year later apartments grew again, with 2014 and 2015 being the turning point for short-term rentals in Seville. The ease of advertising on these platforms and Seville's success as a tourist destination meant that many Sevillians chose to rent entire flats and houses to visitors as a permanent business. Until 2016, most short-term vacation rentals were unregulated, except for those that were officially recognised as apartments. In this year the Andalusian government passed new legislation that enforced administrative control over this business through licenses, so a difference with hotels and especially apartments was made and statistics began. Unlike

**Figure 5.** Continued

apartments, tourist homes do not have to be in a building for tourist purposes, rather they usually are in those with a residential use. A tourist home can be a room in a private house or, as in most cases – up to almost 78% – an entire property. Beds in that accommodation category have skyrocketed ever since, up to 18,468 in 3,897 tourist homes, whereas there are currently 22,140 beds in 218 hotels and hostels across the city. When all the establishments are added together there are 44,220 beds in Seville, a figure that undoubtedly falls short, as beds in tourist homes only show a partial reality because many are operating without a license.

This issue is readily visible if we compare official and advertised data for tourist homes. DataHippo indicates there were 9,553 tourist homes in the city on Airbnb and HomeAway websites in 2018, although the figure might be lower due to aggregation issues. Yet, as we have said, overtourism is not just about the accommodation offer. Figure 5(a) and 5(b) show respectively the distribution of tourist homes and hotels, as well as their spatial density. Both variables concentrate on Seville's historic district and behave differently. Hotels cluster around the district's southern half, close to the UNESCO sites in Santa Cruz and to the west, in the Arenal and Museo neighbourhoods. Unlike hotels, tourist homes are spread over the historic district, including the northern area, with a huge concentration in the Santa Cruz, Alfalfa and Alameda

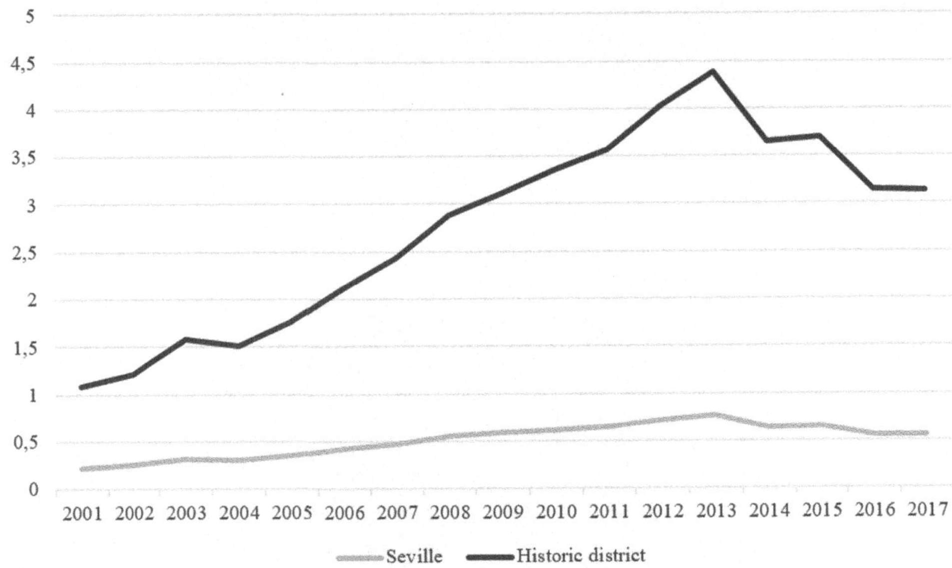


**Figure 6.** Top 5 OECD nationalities and their relative weight in Seville and its historic district.
Source: Authors from Seville City Council Statistics Service.

quarters. Arenal and Museo are also popular, as is Triana across the river. The increase in the offer should not surprise us if we take into account how demand has evolved. According to the Spanish Statistics Office, tourist numbers have grown by 80% in the past decade, with overnight stays reaching 5.5 million in 2018.

Regarding distribution patterns, hotels follow the logic of the southern half of the district being the city centre. Respondents to our survey generally made a strong differentiation too. More precisely, they, especially those who have been residents for longer periods of time, tend to talk about the southern half as being almost entirely specialised in tourism. As has been said, there is a perception of tourism expansion towards new spaces, especially in the northern quarters, where many participants live or used to live. Some are bemused to see groups of visitors entering a part of the city where 'chavs used to hang out' (EB2) or homeless people still gather (EB4). Besides, the feeling is stronger once tourism is no longer confined to particular seasons: 'My perception is that there was tourism before, but only in summer and in the south side. They [the tourists] didn't come here [to the northern quarters]. Now they do. I don't know whether there are more tourists or if they walk further' (EB4).

Most interviewees, especially those who were forced out of the district, point to a 'tourism breakdown' in certain areas. Some (EB1, EB4, EB5 and EB6) link the increase of tourists with the rise in the cost of living. EB2 claims she stopped walking through certain streets where visitors pass by in the five years before she moved out, which she did recently. She also feels that because of tourism, meeting points for people living in the area have decreased, as shops constantly change and public squares are overcrowded. Respondents perceive an increase in dirtiness too, pointing to tourists' behaviour, which in turn could cause locals to care less for their public spaces. Actually, EB4 blames residents for this situation: 'Tourists don't bother me. What does are people who always give priority to foreigners, instead of making easier living conditions for residents'. In this sense, EB6, a French national who has been living in Seville for years and used to reside in the historic district, told us of a recent anecdote: 'A very unpleasant thing happened in a pub that used to be one of my old haunts… One morning I decided to go there on my own to have breakfast. The bill was 3 euros for half a piece of toast and a cup of coffee [very expensive by Seville's standards], and it was because I look like a tourist, so I decided not to go back anymore'.

### *5.3. Lifestyle migration*

Seville's foreign population has increased in the years under scrutiny from 2.3% in 2001 to more than 7% in 2017, despite some deceleration registered in 2015. Most migrants have economic grounds for moving (Torres et al., 2011), but there is also room to consider lifestyle migrants. Migrants from the top 20 Human Development Index nationalities in Seville make up to 3% of the city's total population, while the top 5 most common OECD nationalities make up less than 1%. However, the spatial distribution of these migrants is worth studying. The percentage of Sevillians in the historic district is lower than in the city as a whole, decreasing in the years under scrutiny, whereas numbers of non-Sevillian Spaniards remain stable. By contrast, foreign residents in the district have gone from making up almost 4% of the population in 2001 to 15% today, with significant growth being the result of lifestyle migrants, especially those from the top 5 OECD nationalities (British, French, Germans, Italians and North Americans) who make up 40% of the district's foreign population. In 2001, the most widely represented nationalities in the district came from Latin America, Africa and Asia, but this changed in 2011. For example, numbers of Moroccans fell from 8.1% of foreigners in the district in 2001 to 5.5% in 2011, while Italians went from 6.5% to 10.7% over the same period (data from national censuses, Spanish Statistical Office). Despite this, the top 5 OECD nationalities follow a similar tendency to national residents in the district, growing in the century's first decade to then decrease after 2013. However, they are still relevant and growing in comparison to the city as a whole, as seen in Figure 6.

As for the reasons as to why they settle there, expats living in the area can shed some light. Many recognised they came for leisure or to study and decided to stay in large part due to the historic district, putting forward a cultural argument. For example, EA2, a French national in her thirties, says: 'I wasn't sure why, but I wanted to live here, it has something special, the historic side, the charm, the colour, the light'. Similarly, EA5, a German who has lived in Seville for 38 years, mentions the city attracted him because of 'the love, the magic, the fantasy… the flamenco, its freedom and rebelliousness'. In parallel, there is the idea of a community feel that previous respondents also brought up. EA1, a 40 year-old Italian who settled in the northern side of the district in 2004, says 'I love to hit the street knowing something may come up. In Italy you meet up with friends and have to be punctual, here you don't arrange anything, you come across people'. EA2 also comments on the district's northern side: 'I like the "barrio" atmosphere, the little shops, the friendly environment where people know each other'. EA4 and EA6 share similar reflections, although they have moved in recently (5 and 2 years ago respectively), so their opinions differ. EA4, a retiree from the Netherlands who sold his house in his homeland to buy and restore an old building in Seville that he now rents partially on Airbnb, highlights: 'I enjoy the streets, the bars, everything is cheap and authentic'. EA6, an Italian in her early thirties who works as a hotel receptionist, mentions she often enjoys being around fellow countrymen: 'My favourite place is in Alameda … When I want something familiar or I want to relax, I go to a café there. I'm friends with the owner, he's Italian too'. Most participants have friends from their own nationality living nearby, highlighting it as another reason why they like the district besides proximity to their jobs or to services. EA6 is not the only interviewee working in the tourist sector; EA2 is tourist guide, while EA3 has a restaurant precisely in Alameda quarter whose clientele is increasingly foreign. EA1 recognised she has also thought about changing jobs to do something related to tourism, although her testimony is relevant in another direction. She commented that her main reason for settling here was cheap rents, highlighting how prices have risen:

> 'I lived close to Alameda for 10 years (between 2004 and 2014). Rent was €370 for 70 square metres. Then I moved to another place and I paid €400 for 50 square metres … Recently I found a tiny attic for €250, but I had to move again. When the owner said I had to go it didn't surprise me, because he had refurbished the floor below me and turned one house into two. One of the builders told me it was for holiday rentals. Other residents in the building didn't agree [with what he was doing].'

EA5 confirms he and his colleagues, many of them also expats, have noticed this process, and reflects on it: 'I guess everybody looks for the same: the picturesque, the beauty, the lifestyle in the streets too. That's what the historic district had, especially before. Now it is another thing, closer to a theme park'. It is notable that all participants, to different degrees, have noticed urban transformations in recent years. Even EA6, who only arrived recently but did her Erasmus in Seville ten years ago, has experienced it: 'Prices have gone up, you can't find flats for workers or students, it's all for tourists … Before you could choose between 10 flats with more or less a similar rent, but now you only have 3 apartments in bad shape and really overpriced'. Most respondents also recognised that they deliberately avoid certain areas where tourism is more intense, around the UNESCO sites, and would only go there when they have visitors. EA5 insisted: 'Tourism is necessary and good, but it has become a servitude. Any new store that opens is another franchise, another bar, and another souvenir shop. There's no alternative here'. EA1 and EA6 shared this thought, with EA6 commenting:

> Thanks to tourism I can make a living in Seville. But there are bad things, such as the higher rents and the erasure of traditional places. For example, there are many hipster bars opening … In town you just find inauthentic places now. Which isn't wrong at all, I like to have a *frappucchino* sometimes, but I remember when there was one Starbucks, now there are so many!

### *5.4. Displacement and social unsustainability*

There are two key ideas that local and foreign residents generally perceive. Firstly, there is the production of an urban community or 'barrio' in the successive processes of settlement in the historic district, despite – and indeed sometimes as part of

–gentrification. Nevertheless, many interviewees describe these communities as ephemeral, as people, businesses and places have been constantly changing for the past two decades. In fact, many feel a sort of distance and resentment towards the area. Related to this, there is also a rejection of overcrowding of people in the streets, shops and pubs, as well as road traffic congestion. For example, EB3 was abroad for some time, and when she returned to Seville: 'I didn't want to come back to the city centre, because the "barrio" I knew wasn't there anymore'. The feeling of some kind of displacement is strong, with different views being adopted. Some consider themselves at a disadvantage due to the rise in the cost of living, both in terms of rents and basic goods, repeating the feeling of estrangement from the place – often even if they still reside in the area. Others have a similar perception, but have moved away because they wanted to buy instead of rent. For instance, EB6 comments that 'we decided to buy a flat because we wouldn't afford rising rents in the city centre', while EA2, who is looking for a flat at the moment, links this increase in rents to short-term rentals:

> There are way more vacation homes, I barely have neighbours now. The building next to mine is entirely for tourists. In my building there are four [short-term rentals] … I liked going on holidays knowing the person across the corridor has my number in case anything happened. Now I never know who's there. That bothers me… When I'm at work, tourists ask me how much I pay and where I live. Then they tell me they are staying in a beautiful Airbnb, and they aren't aware that a family used to live there before paying a normal rent, and now they can't because it has risen too much.

EA2 has also noticed how supermarket franchises targeting tourists have increased. In general, there is a perception of displacement of traditional retail businesses (local shops and pubs) in favour of franchises and large firms. For instance, EB4 states: 'With time, there weren't local shops anymore, just franchises, even the market [*Mercado de Feria*, close to Alameda] had become expensive… many greengrocers and fishmongers closed… even the hardware store closed!'. In line with land uses within the district, many participants also make a significant differentiation between quarters. EB1 claims to have changed the area where she used to spend time with friends progressively from west to east, before leaving the district. She was first evicted from San Lorenzo and moved eastwards to Alameda, but soon after she felt that nightlife and tourism had taken over the area so decided to leave. Likewise, EB2 and EB4 moved from Alameda to San Julián (in the district's north-east corner) before moving out. EA1 and EA6 have followed a similar path but managed to remain in the district.

Taken together, interviewees' reasons for moving towards other areas are the rise in the cost of living, where house prices are fundamental, and more broadly to escape from the impact of tourism on the socio-cultural fabric. It is significant that among both current and former residents there is a feeling of disorientation that is the opposite of the sense of community that attracted many of them to the area at first. 'Before, you knew the bartenders, the pub owners, your neighbours… Now you don't know if they are from Madrid or China. It changes a lot, pubs are open for two years and they change again' (EB1). EA1 and EA2 share that view, and the latter adds: 'If tourists ask me… I'd rather recommend tourist quarters like Santa Cruz, because it's already ruined, than suggest they go to other traditional "barrios" in the district'. EB3 states:

> The district is depersonalized. It changes every six months, shops and people, so is harder and harder to make social bonds ... Only homeowners stay, you can't stay renting nowadays... My feeling is the city centre is not a "barrio" anymore. You don't know your neighbours because of vacation homes. Before there were more people living in the area, living there all year round.

EC1 and EC2 share this perception, but not all current residents agree here. For instance, EC4, a professional who moved in at the beginning of the 1990s, likes the transformation of the built environment and the social milieu around Alameda. In his opinion, the square's renewal has made the space available to everybody in the city, whereas before it was appropriated by very specific social groups. The area has changed from a marginal place with a strong underground culture to a metropolitan quarter for Seville's entire population and its visitors. A few current and former residents share this opinion (EA5, EB1 and EC3). Similarly, those who have settled down recently also have a positive view about these particular upscaling transformations and do not feel distance or resentment as long-standing residents do. This would be the cases of EA6 or EC5, a young professional who moved in recently and views the cultural activities in the area positively, perceiving a sense of community.

## 6. Discussion and conclusion

As we have shown, the population of Seville's historic district declines from 2012 onwards, with this being a turning point compared to previous tendencies. This reduction corresponds to a decrease in adult households, precisely a group that had been growing since the 1990s. A possible explanation would be that a transnational gentrification process is taking place, where local low and middle classes are expelled from the district as a consequence of rising rents and cost of living, being replaced by lifestyle migrants with a higher purchasing power. However, demographic data and expats' testimonies illustrate how foreign households from richer countries may also struggle and, in some cases, are being displaced in a similar way to the local population in the area. All lifestyle migrants interviewed decided to settle in Seville's historic district for cultural- and/or leisure-related reasons, yet not all enjoy a higher purchasing power, which could be assumed because of their nationality. Therefore in a few cases their behaviour in terms of where to work – for instance in the tourist sector – and to live shares similarities with that of Sevillians. Thus, there is not enough evidence to sustain that a process of transnational gentrification is active. Lifestyle migrants' consumer practices may be intertwined in many cases with those of tourists, and their role in attracting tourism could be worth studying in-depth, but they are not at the core of the urban transformations and increasing social unsustainability taking place in Seville's historic district. Lifestyle migrants cannot be strictly grouped with tourists when it comes to analysing urban processes, as they may experience the consequences of touristification too.

The historic district's population decline since 2012 coincides with the recovery after the economic crisis, and more specifically, with an increase in tourism numbers, following strategies that had been planned out in some cases even before the crisis. As we have seen, these increasing numbers correspond to the accommodation offer, especially the boom in short-term vacation rentals, and to the demand, which has grown at the same time as permanent residents (local or foreign) have been decreasing in number. Therefore, an alternative explanation to the process is that tourists are competing for space – mainly in the housing market – with residents, forcing out those with less income. This explanation coincides with the perception of both former and most current residents of the historic district that we interviewed, who highlight the ever-changing quarters, where every season there are new people and a growing number of trendy shops. As a result, the character of the 'barrio' vanishes and even perceived 'authenticity' disappears, which in local cultural codes means the absence of anything resembling a community. Previous processes of gentrification in Seville's historic district were ethically questionable and profoundly unjust, as there is evidence of the displacement of local communities since the early 1990s. Nevertheless, local or foreign middle classes settling in the area were at least creating new communities, a possibility that current urban processes make more difficult nowadays.

The high specialisation in tourism and its distribution towards traditionally residential quarters is culminating in overtourism of those areas instead of alleviating already saturated hotspots. By drawing on demographic and tourism statistics combined with the analysis of residents' perceptions, we have shown that Seville's historic district is suffering from overtourism, with there being differences between neighbourhoods such as Santa Cruz, Alfalfa or Alameda. Figures may not be as high as in other urban Mediterranean destinations like Barcelona or Venice, but the effects on the socio-cultural fabric share similarities due to a strong spatial concentration. Understood as an increase in tourist activity that reduces the attractiveness of the conditions that precisely make tourism possible, overtourism is affecting residents, both nationals and foreigners, who increasingly avoid the most touristy areas or are disappointed because the neighbourhood they moved to has changed radically. Furthermore, many of these residents cannot afford to live in the historic district due to touristification, as a few of our participants have recognised to differing extents. All this together is not only making the reproduction of the already shrinking traditional local communities impossible, but it also makes the development of new stable communities unlikely. Seville's historic district is socially unsustainable as these processes produce a city that is only for visitors and that has come to reject many of those who actually live in it and shape it every day. However, further research on the topic is needed to establish a correlation between displacement and increasing tourism or to deepen our knowledge of how overtourism affects visitors' perceptions, as well as more studies on a micro scale, focusing on particular neighbourhoods or even on significant streets.

## Acknowledgment

We would like to thank two anonymous referees for their insightful comments on a previous draft of this paper. The usual disclaimers apply.

## Disclosure statement

No potential conflict of interest was reported by the authors.

## Funding

We acknowledge support from the Portuguese Science and Technology Research Council (FCT) within the scope of the SMARTOUR project (Ref: PTDC/GES-URB/30551/2017) and the Spanish Ministry of Science and Innovation through the LIKEALOCAL project (Ref: RTI2018-093479-A-I00).

## References

Annunziata, S., & Lees, L. (2016). Resisting austerity gentrification and displacement in southern Europe. *Sociological Research Online, 21*(3), 148–155. doi:10.5153/sro.4033

Ashworth, G., & Page, S. (2011). Urban tourism research: Recent progress and current paradoxes. *Tourism Management, 32*(1), 1–15. doi:10.1016/j.tourman.2010.02.002

Assemblea de Barris per un Turisme Sostenible. (2019, 8 July). *Hipermovilidad, turistización, cambio climático y redes de auto-organización*. Retrieved from https://assembleabarris.wordpress.com/2019/07/08/hipermovilidad-turistizacion-cambio-climatico-y-redes-de-auto-organizacion/

Benson, M. (2012). How cultural significant imaginings are translated into lifestyle migration. *Journal of Ethnic and Migration Studies, 38*(10), 1681–1696. doi:10.1080/1369183X.2012.711067

Bianchi, R. (2018). The political economy of tourism development: A critical review. *Annals of Tourism Research, 70*, 88–102. doi:10.1016/j.annals.2017.08.005

Blanco-Romero, A., Blázquez, M., & Cànoves, G. (2018). Barcelona, housing rental bubble in a tourist city. Social responses and local policies. Sustainability, *10*(6), 2043–2018. doi:10.3390/su10062043

Blázquez, M., & Murray, I. (2010). Una geohistoria de la turistización de las islas Baleares. *Periplo Sustentable, 18*, 69–118.

Calle Vaquero, M. (2006). *La ciudad histórica como destino turístico*. Barcelona: Ariel.

Cañada, E. (2018). Too precarious to be inclusive? Hotel maid employment in Spain. *Tourism Geographies, 20*(4), 653–674. doi:10.1080/14616688.2018.1437765

Cheer, J. (2018). Geographies of marginalization: Encountering modern slavery in tourism. *Tourism Geographies, 20*(4), 728–732. doi:10.1080/14616688.2018.1477171

Cócola-Gant, A. (2018). Tourism Gentrification. In L. Lees & M. Phillips (Eds.), *Handbook of gentrification studies* (pp. 281–293). Cheltenham and Northampton: Edward Elgar Publishing.

Cruz, J. (1986). *La población de Sevilla*. Seville: Ayuntamiento de Sevilla. doi:10.1086/ahr/84.2.472

Dempsey, N., Bramley, G., Power, S., & Brown, C. (2009). The social dimension of sustainable development: Defining urban social sustainability. *Sustainable Development, 19*(5), 289-300. doi:10.1002/sd.417

Díaz-Parra, I. (2010). *Sevilla, cuestión de clase*. Seville: Atrapasueños.

Díaz-Parra, I. (2014). *¿Gentrificación o barbarie? Disciplinamiento y transformación social del barrio de la Alameda de Sevilla*. Seville: Atrapasueños.

Díaz-Parra, I. (2016). Sevilla 1929-1992. La producción de una mercancía. In La Corrala (Ed.), *Cartografía de la ciudad capitalista* (pp. 195–201). Madrid: Traficantes de Sueños.
Eramo, M. (2014). Unescocide. *New Left Review, 88*, 47–53.
Goodwin, H. (2017, October). The challenge of overtourism. Responsible Tourism Partnership (Working Paper 4). Retrived from https://haroldgoodwin.info/pubs/RTP%27WP4Overtourism01%272017.pdf.
Gotham, K. (2005). Tourism gentrification: The case of New Orleans' Vieux Carre. *Urban Studies, 42*(7), 1099–1121. doi:10.1080/00420980500120881
Hayes, M. (2018). *Gringolandia: Lifestyle migration under late capitalism*. Minneapolis: University of Minnesota Press.
Huete, R., & Mantecón, A. (2018). El auge de la turismofobia ¿hipótesis de investigación o ruido ideológico? *PASOS, 16*(1), 9–19.
Jover, J. (2019). Geografía comercial de los centros históricos: entre la gentrificación y la patrimonialización. El caso de Sevilla. *Boletín de la Asociación de Geógrafos Españoles, 82*(2788), 1–33. doi: 10.21138/bage.2788
Jover, J., Berraquero-Díaz, L., Barrero-Rescalvo, M., & Jiménez, A. (2018). Turistización y movimientos urbanos de resistencia: Experiencias desde Sevilla. In C. Milano & J. Mansilla (Eds.), *Ciudad de vacaciones* (pp. 403–437). Barcelona: Pol.len.
Jover, J., & Díaz-Parra, I. (2019). Gentrification, transnational gentrification and touristification in Seville, Spain. *Urban Studies*. doi:10.1177/0042098019857585
Koens, K., Postma, A., & Papp, B. (2018). Is overtourism overused? Understanding the impact of tourism in a city of context. Sustainability, *10*(12), 4384. doi:10.3390/su10124384
Milano, C. (2018). Overtourism, malestar social y turismofobia. Un debate controvertido. *Pasos. Revista de Turismo y Patrimonio Cultural, 16*(3), 551–564. doi:10.25145/j.pasos.2018.16.041
Murray, I. (2015). *Capitalismo y turismo en España*. Barcelona: Alba Sud.
Polese, M., & Stren, R. (2000). *The social sustainability of cities. Diversity and the management of change*. Toronto: University of Toronto Press.
Richards, G. (1996). Production and consumption of European cultural tourism. *Annals of Tourism Research, 23*(2), 261–283. doi:10.1016/0160-7383(95)00063-1
Sequera, J., & Nofre, J. (2018). Shaken, not stirred. New debates on touristification and the limits of gentrification. City, *22*(5-6), 843–855. doi:10.1080/13604813.2018.1548819
Seraphin, H., Sheeran, P., & Pilato, M. (2018). Over-tourism and the fall of Venice as a destination. *Journal of Destination Marketing & Management, 9*, 374–376. doi:10.1016/j.jdmm.2018.01.011
Seville City Council. (2003). *Plan Estratégico Sevilla 2010*. Seville: Ayuntamiento Sevilla.
Seville City Council. (2006). *Plan general de Ordenación Urbana*. Seville: Ayuntamiento Sevilla.
Sigler, T., & Wachsmuth, D. (2015). Transnational gentrification: Globalisation and neighbourhood change in Panama's Casco Antiguo. *Urban Studies, 53*(4), 705–722. doi:10.1177/0042098014568070
Smith, N. (1996). *The new urban frontier. Gentrification and the revanchist city*. London: Routledge.
Smith, N. (2002). New globalism, new urbanism: Gentrification as global urban strategy. Antipode, *34*(3), 427–450. doi:10.1111/1467-8330.00249
Torres, F.J., Galera, V., Fernández Salinas, V., Brivio, F., Chioua, J., … González, A. (2011). *El distrito Macarena de Sevilla. Migraciones recientes y transformaciones urbanas*. Seville: Junta de Andalucía.
UNWTO. (2018a). *Overtourism? Understanding and managing urban tourism growth beyond perceptions. Executive summary*. Retrieved from https://www.e-unwto.org/doi/pdf/10.18111/9789284419876.
UNWTO. (2018b). *Tourism highlights, 2018 edition*. Retrieved from https://www.e-unwto.org/doi/pdf/10.18111/9789284419876.
Villar Lama, A., & Fernández-Tabales, A. (2017). Reconstruir la historia del turismo a través de la prensa: La evolución del espacio turístico de Sevilla (1915-2015). *Cuadernos Geográficos, 56*(1), 290–321.

Wachsmuth, D., & Weisler, A. (2018). Airbnb and the rent gap: Gentrification through the sharing economy. *Environment and Planning A: Economy and Space, 50*(6), 1147–1170. doi:10.1177/0308518X18778038

Yrigoy, I. (2016). Financialization of hotel corporations in Spain. *Tourism Geographies, 18*(4), 399–421. doi:10.1080/14616688.2016.1198829

Zukin, S. (1995). *The culture of cities*. Oxford: Blackwell.

# Zoning for world heritage sites: Dual dilemmas in development and demographics

Thomas E. Jones, Huong T. Bui and Katsuhiro Ando

**ABSTRACT**

Since listing as a UNESCO World Heritage Site (WHS) in 1999, Hoi An Ancient Town in Vietnam has been transformed by tourism development with visitor numbers increasing more than sixteen-fold. Beyond positive economic impacts, tourism has generated considerable funds for conservation, with revenue assisting authentic restoration of the town's historic center. However, tourism impacts have provoked criticism linked to social sustainability, examined in this paper using the core-and-buffer-zone principle - a neo-normative spatial planning framework. Our longitudinal, mixed-method analysis from development and demographic perspectives combines interviews and census data with spatial planning guidelines and maps. Findings show that architectural renovations, accompanied by stricter regulations and hierarchical World Heritage Site zoning, legitimized rapid development of Hoi An's buffer zone in tandem with 'museumification' of the core, epitomized by the central clustering of traditional wooden architecture renovation projects. By 2010, population in the core had declined by 20% as 'adaptive re-use' of heritage buildings paved the way for conversion of private residences into shops or other services and the core transformed from living heritage into a tourism stage. Demographic and development trends reinforced the dual hierarchy, although attempts to mitigate museumification, including a series of craft villages around the perimeter, provided impetus for revitalization of the newly-expanded buffer zone. Our findings highlight certain implications of spatial zoning for social sustainability, pointing the way for better integration with adaptive re-use policies. Our study contributes to further the debate on social sustainability at a living heritage site by investigating the bipolarity between museumification of the core and concurrent development of the buffer zones. The implications of our study extend beyond the current context of developing Southeast Asia, as this paper draws parallels and expounds opportunities for more site-specific planning and management of World Heritage Sites.

**摘要**

越南的会安古城自1999年被联合国教科文组织列为世界文化遗产以来，已经被旅游发展所改变，游客数量增加了16倍之多。除了积极的经济影响外，旅游业也为保护工作带来了可观的资金，收入有助于该城历史中心的真正修复。然而, 旅游影响引发了社会可持续性方面的批评, 本文使用新规范空间规范规划框架中的核心-缓冲区原理对之进行了研究。我们的纵向、混合方法分析, 采取发展和人口统计多个角度, 综合了访谈、人口普查、空间规划指南和地图多源数据。研究结果表明, 建筑更新伴随着更严格的规定, 同时该世界文化遗产的分级分区治理体系, 赋予会安古城核心区的 “博物馆化”的同时缓冲区的同步快速发展, 体现了传统木结构建筑更新项目核心分类治理的体系。到2010年, 随着遗产建筑的”适应性再利用”为将私人住宅转变为商店或其他服务设施铺平了道路, 核心区也从生活性遗产阶段转型为旅游遗产阶段, 核心区域的人口减少了20%。该世界遗产的人口和发展趋势加强了不同区域之间的等级关系，尽管减轻博物馆化的尝试（包括周边的一系列手工艺品村庄）为振兴新扩展的缓冲区提供了动力。我们的研究通过调查核心地带的博物馆化和缓冲区的同步发展之间的两极化, 有助于进一步讨论活态遗产的社会可持续性。我们的研究的意义超出了目前东南亚发展的背景, 因为这篇论文利用世界文化遗产的共性, 为更多世界文化遗产的场所管理与规划阐述了多种机遇。

## Introduction

Prior to the coronavirus pandemic, an era of unparalleled mobility had transformed Southeast Asia into a stage for unprecedented growth in tourism demand. However, concerns for heritage conservation have arisen due to transformation of physical landscapes, coupled with gentrification issues that have left residents disconnected from their past and their sense of place (Winter & Daly, 2012). Despite the common management issues that converge on UNESCO World Heritage Sites (WHSs), such negative impacts tend to be eclipsed by the attractiveness of listed status and its purported 'tourism enhancing effect' (Yang et al., 2010). The post-millennial inscription and rapid development of Southeast Asian WHSs testifies to many such positive impacts including the contribution of WHS tourism to GDP growth, job creation and poverty alleviation. However, a longer-term assessment also reveals less-alluring socio-economic consequences of WHS listing including 'social polarisation' (Jimura, 2011), exacerbated by hierarchical Eurocentric planning guidelines that favor static archaeological parks or museums over intangible heritage. This paper investigates one such UNESCO-endorsed spatial planning framework, the core-and-buffer-zone principle, to examine social sustainability from the perspectives of demography and tourism development in Hoi An – a living cultural WHS located in Quang Nam Province, central Vietnam. By leveraging the increased international profile that accompanied WHS listing in 1999, Hoi An has become an archetypal Southeast Asian WHS 'success story' (Brooks, 2008). A plethora of international awards and sixteen-fold increase in tourist numbers support this positivist view of development (Di Giovine, 2009a), but at the other extreme some see the WHS status as a 'death sentence' for the ancient town (Avieli, 2015), that suffers 'one of the most glaring examples of museumification' (Di Giovine, 2009b). This paper aims to investigate the influx of tourists and exodus of local residents from Hoi An's core zone to assess the long-term social sustainability of WHSs. The research objectives are to examine: (1) longitudinal trends in tourism development and resident demographics;

and (2) the dynamics of museumification in the core and regeneration of the buffer zone.

Our work contributes to prior literature via a long-term perspective of Hoi An that spans pre- and post- WHS inscription, corresponding to calls for site-specific heritage management approaches to complement UNESCO's Eurocentric guidelines (Logan, 2002). A mixed-method approach enables holistic reflection on the tourism-heritage spectrum that includes socio-spatial consequences of the core-and-buffer-zone policy. In particular, the evaluation of spatial zoning is considered inseparable from the concurrent renovation of traditional architecture, and the transformation in socio-economic conditions and lifestyle. First, a literature review critically examines the concepts of social sustainability, WHS management and zoning, along with museumification's positioning within the gentrification nexus. Next, findings are arranged around the trends in development and demography unfolding in Hoi An's core and buffer zones. Parallels are drawn with spatial planning and social sustainability at other Southeast Asian WHSs such as Lijiang and Angkor Watt, and implications drawn for UNESCO planning protocols.

## Literature review

### *Social sustainability in heritage towns*

Heritage management has evolved from monuments to historic towns and cities (Hassan & Xie, 2020). However, the conservation of living landscapes transcends tangible heritage and architectural blueprints to encompass the triple bottom line of economic, environmental and social concerns (Orbasli & Woodward, 2009). Within the sustainable development debate, social sustainability is identified as social equity and the sustainability of communities - the former centers upon a distributive notion of social justice, while the latter relates to the viability, health and functioning of a community (Dempsey et al., 2011, p. 298). Compared to economic and environmental dimensions, social sustainability indicators such as cohesion and inclusiveness are arguably harder to measure, but more important for a destination's long-term durability. Unlike monuments or archaeological sites, or 'set aside' national parks, historic towns still house resident communities that can form as much of the attraction as the physical landscape (Orbasli, 2000). A 'living heritage approach' (Poulios, 2014) is thus required to marry spiritual and intangible beliefs even at the expense of holistic conservation of tangible or material heritage.

However, simultaneously sustaining living heritage for local residents and tourists is a tightrope that depends on various factors such as local involvement, education and training, a balance of authenticity and interpretation, and integrated planning and support policies (Loulanski & Loulanski, 2011). In the absence of such checks and balances, tourism can run counter to conservation (Chapagain, 2013). Although it relies on a common pool of assets akin to cultural conservation, the tourism sector is predominantly concerned with profit margins while heritage management must answer to a broader range of objectives (McKercher & du Cros, 2003). Without robust management, 'modern mass tourism blights the aura of all notable relics' (Lowenthal, 2015, p.

425) and the risk of heritage commodification or gentrification is only heightened at iconic destinations such as UNESCO WHSs.

### *Social sustainability and WHS management*

A recent global backlash in WHS honeypot cities such as Venice and Barcelona has reiterated that tourism is no panacea and careful site management is paramount to stem the spread of residents' 'tourism-phobic' perceptions (Zerva et al, 2019). Although the WHS label can function merely as a placebo rather than a direct economic catalyst (Adie et al., 2018), listing often does have a significant 'tourist enhancing effect' (Yang et al., 2010). However, several factors are found to moderate the effect of WHS status on tourism demand including GDP level of the destination country and heritage type (cultural or natural) (Yang et al., 2019). Nonetheless, the increased scrutiny that accompanies WHS listing invariably exacerbates pressure on social sustainability and site management. The ambiguousness of UNESCO's underlying philosophy of 'outstanding universal value' struggles to offer effective compromises that can balance the priorities of WHS listing with local needs (Su & Wall, 2012). Extra global recognition may also trigger conflicts between local place attachment and ownership and create gentrification stress for residents in and around the WHS (Orbasli, 2000). At an everyday level, the presence of increasing numbers of visitors and tourist-oriented activities impacts on the lives of residents (Liang & Hui, 2016). The cost of new facilities and infrastructure development combined with the top-down, often opaque WHS inscription process itself can also isolate, frustrate or even enrage local communities (Jimura, 2011). Despite the recent spread of 'over-tourism,' it remains hyperbole to talk of 'UNESCO-cide,' wherein it is argued that listed status can become a 'death sentence for a site' (Avieli, 2015). Nonetheless, many WHSs do undergo a process of gentrification via a complex chain reaction of rising land prices, commercialization and out-migration of long-term locals as outsiders move in (Phillips, 2004). Spatial planning tools such as zoning are among the most widely-deployed management tools implemented at WHSs to mitigate these negative impacts.

### *Using core and buffer zones to manage WHSs*

The 'core-and-buffer-zone' principle corresponds to a Euclidean model of territorial distribution, a nomenclature derived from a 1926 US Supreme Court decision in Euclid, Ohio that set a legal precedent for local government zoning (Gillespie, 2012). With the formal recognition of town planning jurisdiction, tiered territorial hierarchies proliferated across heritage fields and buffer zones were set up to protect the core. Zoning systems spread via 'cross-pollination' whereby ecologist experts tasked with inspection of *natural* WHSs dispersed the philosophy to their *cultural* heritage equivalents.

The concept of buffer zones dates back to the Operational Guidelines for the implementation of the World Heritage Convention in 1977. The 2002 Budapest Declaration formalized the transition from conservation to heritage management (Hassan & Xie, 2020) and the submission of a WHS management plan became a *de facto* requirement. With the apparent endorsement of UNESCO, zoning became ubiquitous as a neo-normative planning tool, but as of 2005 the inclusion of a buffer zone into a nomination of a WHS was 'strongly recommended but not mandatory' (UNESCO, 2008c).

Yet evidence of effective zoning at WHSs remains inconclusive as outstanding issues range from contested territorial delineation during the planning stage to socio-spatial consequences upon implementation. Within the inscription dossier, the buffer zone is not deemed to be part of the property *per se*, hence '[d]efining the perimeter and the adjacent buffer of the property becomes a complex process, with the potential for enormous social impacts' (Gillespie, 2012 p.198). In extreme cases, spatial zoning could legitimize eviction or exacerbate social streaming as at Huangshan in China where five radial towns in the buffer zone were found to be excluded from decision-making, and 'problems exported out to surrounding inadequately managed villages and buffer zones' (Xu, Zhu & Bao, 2016). The risk of bipolar development will be discussed later at Hoi An, but even within the core zone, socio-spatial consequences such as gentrification are often apparent (Miura, 2011).

### *Social consequences: gentrification and museumification*

The lack of consensus over the parameters of 'gentrification' reflects the term's conceptual ambiguity across contexts that range from waterfront areas (Hoyle, 1988); suburban (Badcock, 2001); new-builds (Davidson & Lees, 2005, 2009; He, 2009); and student villages (Hubbard, 2008; Smith & Holt, 2007). Gentrification is not confined to urban areas but includes rural regions (Hines, 2012; Phillips, 2004, 2010) and historic centers or market towns (Ashworth & Tunbridge, 2000). Research into tourism-related 'museumification' as a sub-section of gentrification remains relatively sparse (Gotham, 2005). When heritage is rapidly commercialized for tourism, museumification of core zone buildings and landscapes can occur as in cases from Angkor Watt to the historic colonial quarter in Singapore (Miura, 2011).

Gentrification definitions pivot around a process of upward class formation (Butler & Lees, 2006), often characterized by a corresponding reorganization of physical space into affluent enclaves or resorts (Liang & Bao, 2015). Rising property prices and living costs encourage residents to move out and relocate in less costly and congested areas, often located in the buffer zone or further out. Commodification of heritage leads to the creation of landscapes which are described as 'elitist' and removed from the everyday experiences of the local residents, causing the displacement of low-income groups by an incoming 'gentry' (Davidson & Lees, 2009)

Based largely on empirical evidence from Europe and North America (Gotham, 2005), gentrification has been portrayed as a role-reversal of suburbanization, a 'back to the city' movement of newly-wealthy residents (Smith, 2002; Butler & Lees, 2006). However, where Western research has tended to document largely benign styles of gradual displacement (Hackworth, 2002), looser governance in LDC contexts has opened the door to more aggressive models of gentrification, spearheaded by joint public-private ventures that result in the 'extensive relocation of long-term residents away from the [hub] to the peripheries' (Chan et al, 2016). In Nanluoguxiang hutong of Bejing, tourism-oriented landscape renovation has seen conservation goals packaged with extensive real-estate investment (Shin, 2010) reminiscent of the corporate gentrification that links property investment, mortgage brokers and large corporate developers (Hackworth, 2002). Local communities displaced from WHSs in China adopt tourism in lieu of traditional production methods, even though it tends to offer short-term economic benefits while overlooking the concerns and long-term social sustainability of the locals (Su, Wall & Xu, 2016). Elsewhere, local populations have even been physically evicted during the listing process of certain WHSs such as Borobudur in Indonesia and Sukhothai in Thailand (Poulios 2011; Nagaoka, 2016). This displacement reflects the 'hidden cost' of fast-track WHS development trajectory that puts 'unbearable pressures on paradise' (Reeves & Long, 2011, p.3).

Amidst unprecedented tourism growth in Southeast Asia, how do WHSs such as Angkor Watt, Lijiang, Luang Prabang, and Georgetown Penang mitigate negative social consequences of gentrification? In Georgetown, the community involvement and attachment, combined with pride and attitude to WHS designation soon after listing were said to offer strong support for conservation and development (Rasoolimanesh et al, 2017). However, Angkor Watt - once heralded as an example of WHS 'best practice' - has attracted criticism to the extent of 'UNESCO-cide' wherein sites are 'loved to death' (Wescott & Molinski, 1993). Angkor's ZEMP (Zoning & Environment Management Plan) system contributed significantly to the over-emphasis on the 'static landscape as an ancient archaeological park' (Miura, 2011, p.144). Likewise in Lijiang, 'within the core of the ancient town which is subject to strict heritage conservation rules, it is estimated that only a few thousand Naxi are left since most of the residential houses have been appropriated for tourism or have been bought over by the majority Han people who have come to Lijiang to start businesses or for habitation' (Su & Teo, 2008, p.154). At Hoi An, a mixed methodology was employed to investigate the complex relationship between zoning, gentrification and social sustainability in the context of a developing Southeast Asian WHS.

## Methodology

Our longitudinal case study investigated Hoi An's demography and development trends pre- and post- inscription as a WHS. In keeping with the holistic objectives of living heritage conservation, the empirical approach combined various methods and data sources, seeking out cross-cutting and competing forms of evidence to elucidate more convincing findings (Yin, 2009). Sources were drawn from stakeholder interviews and researchers' long-term observations, coupled with socio-demographic and cartographic data. This mixed method approach fills a gap in the tourism-conservation literature, wherein surveys are often conducted in isolation without due reflection of holistic town planning principles (Ashworth & Tunbridge, 2000). At the intra-case study level, it covers the methodological gaps of previous research on Hoi An, which either take a cross-sectional view of the city's tourism (Trinh et al., 2014); investigate alternative frameworks such as social-ecological systems (Bui et al, 2020); or pivot on a particular socio-economic issue (Bui & Le, 2018).

Primary data for this paper were drawn from 22 interviews with a cross-section of stakeholders selected using snowball sampling. Interviewees included government officers; international consultants and NGO staff; business owners; local residents and

**Table 1.** List of interviewees.

| Actor | Organisation | Code/Date | Place |
|---|---|---|---|
| NGO officer | UNESCO Hanoi Office | R1 (Mar. 2015); R8 (Aug. 2015) | Hanoi |
| Municipal government officer (*tourism-related sectors*) | Hoi An Department of Trade and Tourism | R2 (Mar. 2015); R9, R10 (Aug. 2015) | Hoi An |
| International consultant* | PUP, JICA | R4, R5 (Mar. 2015) | Online |
| Local government officer (*conservation-related sectors*) | Hoi An Center for Culture & Sport; Hoi An Center for Monument Management and Preservation; Cham Island Marine Protected Area | R3, R6 (Mar. 2015); R7 (Aug. 2015) | Hoi An |
| Municipal government officer (*administrative sector*) | Hoi An People's Committee | R11 (Aug. 2016) | Da Nang |
| Provincial government officer | Quang Nam Tourism Promotion Center | R12, R13 (Aug. 2016) | Tam Ky |
| Private business manager | Manager of local hotels | R14, R15 (Aug. 2016) | Hoi An |
| Local resident | Local residents from Minh An, Thanh Nam, Cam Thanh, Cam Chau and Cam Nam wards | R16, R17, R18, R19, R20 (Aug. 2016) | Hoi An |
| Researcher* | University and vocational college | R21, R22 (Mar. 2017) | Da Nang |

*denotes interviews conducted in English.
*Source: Authors' compilation.*

researchers. Interviews were conducted mostly in Vietnamese, but English was also used with NGOs or international consultants. Depending on informants' availability, some interviews continued over several sittings not only in Hoi An and Da Nang but also in Hanoi or online (Table 1). Researchers also conducted numerous informal talks with local residents and officers during the two-year fieldwork period from March 2015 to September 2017. The rapport-building and trusting relationships vital for this kind of ethnographic fieldwork are evidenced by the authors' lengthy experience in Hoi An, where one researcher worked in the tourism industry for a total of five years, and another as a consultant for two years. The research team have been involved and witnessed first-hand the renovation of Hoi An since the early 1990s, enabling a long-term investigative case study (Yin, 2009). Although the authors were no longer living in Hoi An by 2015, frequent field visits enabled ongoing access to their informant network including partners in development projects in which they had been directly involved.

Residents, experts and government officials were questioned about their perception of tourism development impacts on various aspects of town planning and resident adaptability to policy and demographic changes. The interview recordings were transcribed and analyzed in conjunction with field notes. Primary data collected from fieldwork interviews, informal talks, observations, field notes and pictures were input into NVivo 11. The conceptual categories were determined prior to reviewing the text (Guest, MacQueen & Namey, 2012). In this applied thematic coding method, nodes and categories that emerged from analysis were arranged spatially according to the principle of 'core' and 'buffer' zones. These included the tourism-induced effects of museumification and regeneration, together with patterns of tourism development and demographic change in areas with more or less restrictions.

Findings that emerged from interviews, informal talks and observations were triangulated with secondary data comprised of statistics collected by government

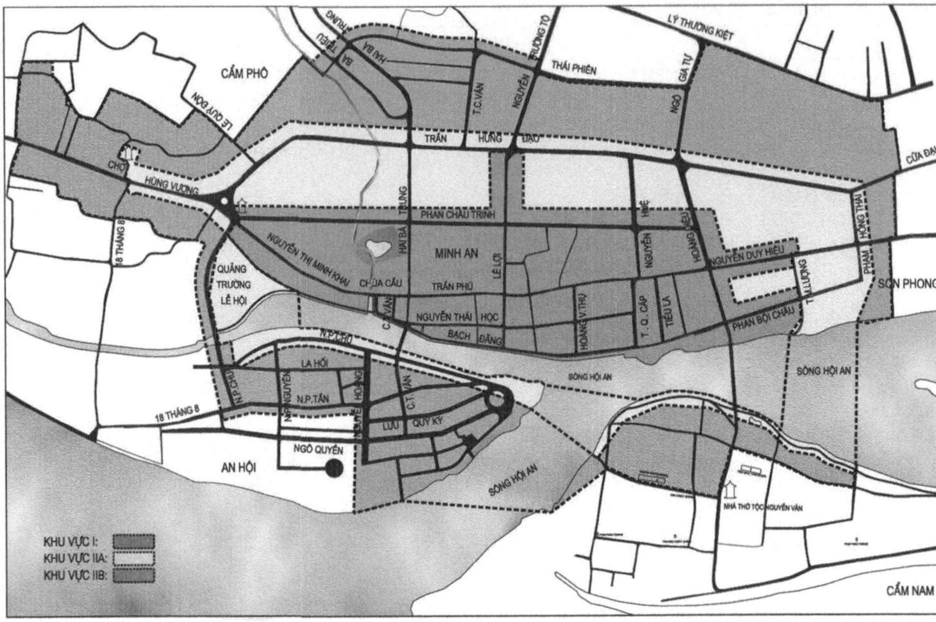


**Figure 1.** Core and dual buffer zoning scheme for Hoi An in 2008.
*Source: Hoi An Center for Monuments Management and Preservation.*

departments together with local and regional masterplans and policy papers. Data was input into Geographical Information System (GIS) with MapInfo software to generate visual depictions of the expanded buffer zone (Figure 1) and population change in different wards (Figure 4). In addition, reports and documents were collected from international non-government organizations (NGOs) and multilateral organizations including UNESCO. Population and employment data from the national household censuses of 2000 and 2010 was combined with Hoi An statistical year books that report annual visitors arrivals, revenue and hotel rooms. Academic and industry publications including theses and reports were analyzed using critical discourse analysis to explore the dialectic link between rhetoric and reality (Fairclough, 2003). Additional data triangulation from different groups of heritage stakeholders, and primary and secondary sources enabled an extra layer of validation (Yin, 2009).

## Zoning for conservation of Hoi An ancient town

Hoi An Ancient Town is located in Vietnam's coastal region of Quang Nam Province, 30 km outside of Da Nang that lies midway between Hanoi to the north and Ho Chi Minh to the south. Once the trade hub of central Vietnam, by the late 18$^{th}$ century Hoi An's significance as a harbour-gateway town had diminished due to changes in the political and hydro-morphological landscapes. Just as Hoi An port's pivotal geo-political standing was buried gently by the gradual siltation of the Thu Bon River estuary, so the ancient town was fossilized, a naturally-preserved backwater (Parnwell, 2016).

Despite the 'stunning outward ossification' of the town today (Di Giovine, 2009a), conservation policy in Hoi An has had to overcome several major hurdles. Following reunification in 1975, the management of some 200 vacant houses abandoned during the Vietnam War was assumed by a public agency (Ando, Fukukawa & Tomoda, 2015). From 1983, conservation of the historic core was driven by Polish architect Kazik who raised international awareness of the traditional wooden architecture of central Minh An Ward that offered a rare window back in time to 'old Ha Noi' (Di Giovine, 2009a, p. 263). Immediately prior to the *Đổi Mới* economic reforms initiated in 1986, Hoi An was designated as National Cultural Heritage and a conservation plan drawn up by local government in 1987 was among the first to delineate the boundaries of the heritage core zone. In 1997, just prior to WHS listing, the second conservation plan formally delineated the core and buffer zones (Figure 1), which continue to underpin contemporary site management (Ando et al., 2015).

In 1999, Hoi An was inscribed as a UNESCO World Cultural Heritage Site for Criterion (ii): 'an outstanding material manifestation of the fusion of cultures over time in an international commercial port' and Criterion (v): 'an exceptionally well-preserved example of a traditional Asian trading port'. After WHS inscription, the 30 ha core zone remained territorially continuous, albeit renamed by the 2008 city master plan as Zone I (the 'intact protected zone'). However, the surrounding buffer zone, covering an area of 280 ha, was divided into Zones IIA and IIB, for ecological and landscape protection purposes. Zone IIA formed an inner ring around the core, including the south bank waterfront, while IIB provided an external ring of protection. The landscape conservation hierarchy was implemented via tiered planning permission laws – for example Zone IIA stipulates a height of 10.5 meters from pavement to roof, and buildings that 'should not have more than two stories,' whereas Zone IIB limits are more relaxed (13.5 meters height and maximum 3 stories) (UNESCO, 2008a). Although the donut-shaped zoning system has remained virtually intact since 1987, the current 2020-2030 master plan envisages an enlargement of the buffer zone across the river-front areas of Cam Nam and Cam Kim, and a terrestrial area encompassing Cam Pho, Son Phong, Tan An and a small section of Cam Chau. Under this new scheme, the total territory of the expanded protection zone would increase threefold compared to the previous benchmark to cover an area of 5.60 km2 (Figure 2), covering 9.2% of the municipality's 60.84 km2 (Prime Minister's Office, 2012).

## Tourism development at Hoi An

While spatial zones have remained mostly unchanged to date, tourism at Hoi An has developed dramatically. In the pre-WHS era, the town was described by long-term observer Avieli (2015) as 'small, easy to navigate on foot… [with] little traffic'. Similar images of an idyllic, rural backwater underpinned early descriptions in a 1991 edition of the *Lonely Planet* guidebook which devoted 14 pages to Hoi An, equal in number to Hanoi or Saigon (Di Giovine, 2009a, p. 263). This image as a 'hidden gem' helped catapult Hoi An to global 'must-see' status via a series of process coincidences that culminated in WHS listing. From an 'off-the-beaten-track' backpacker destination in the early 1990s, Hoi An expanded into a conventional mass tourist destination for trans-Vietnam tours, aided by a convenient location at the heart of the central region's 'cultural corridor' – a trio of UNESCO cultural WHSs (together with My Son and Hue) in

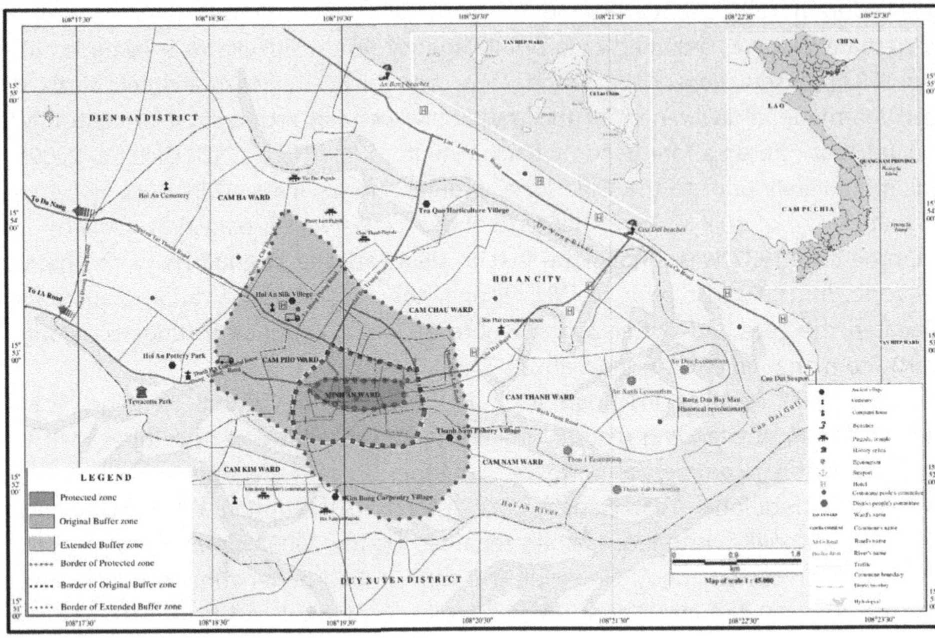


**Figure 2.** Proposal for expanded conservation zone in Hoi An 2020–2030.
*Source: Adapted from* Prime Minister's Office (2012).

**Figure 3.** International and Domestic Tourist Trends in Hoi An (1991–2017).
*Source: Compiled from data of* Hoi An Department of Trade and Tourism Statistical Data (personal communication in 2018).

a 100 km radius around Da Nang (Parnwell, 2016). From pre-WHS 1997, when the total number of tourists stood at 139,982 (including 81,148 international and 58,834 domestic), aggregate tourist arrivals increased more than sixteen-fold to 2,318,241 in 2018 (2,019,685 international and 298,556 domestic). The number of annual overnight stays by international visitors has increased to 1.5 million, with over 2 million international tourists by entrance tickets sold (Figure 3). The domestic equivalent figure stood at

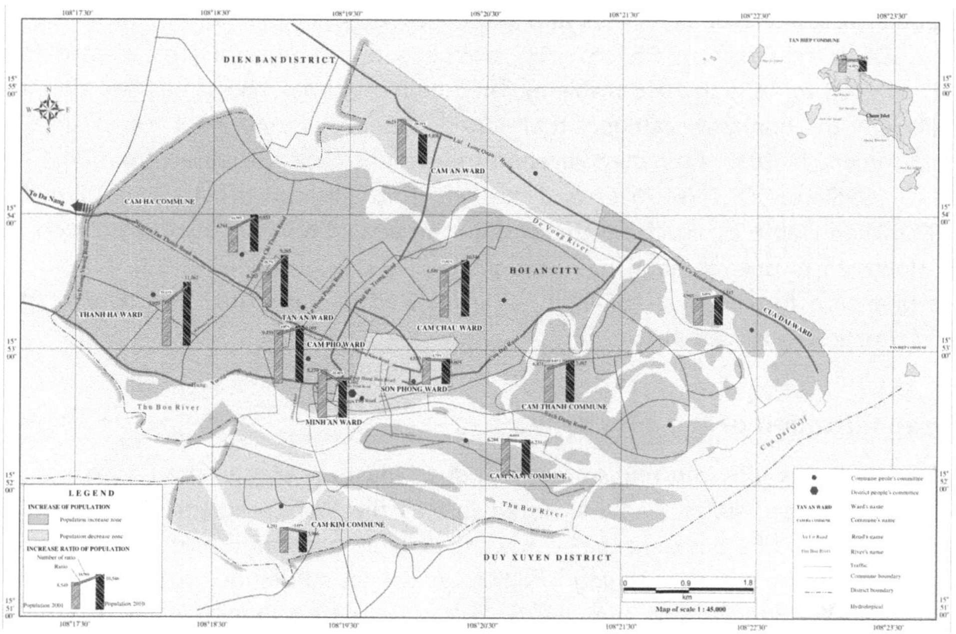


**Figure 4.** Changes in Hoi An's population by ward (2001–2010) Source: Author compilation from data of Hoi An Statistics Office.

**Table 2.** Economic performance of the tourism sector (2007–2016).

| | 2007 | | 2016 | |
|---|---|---|---|---|
| ***Revenue*** | (USD 000) | % | (USD 000) | % |
| Accommodation | 20,073 | 60 | 640,585 | 60 |
| Transport | 816 | 3 | 20,033 | 2 |
| Food & Beverage | 8,973 | 25 | 207,846 | 20 |
| Entrance ticket fees | 1,690 | 5 | 82,131 | 8 |
| Tours | 456 | 1 | 63,548 | 6 |
| Other | 2,391 | 6 | 47,378 | 4 |
| *Total* | *34,399* | | *1,061,522* | |
| **Labour** | (people) | | (people) | |
| | 3411 | | 7535 | |
| **Accommodation sector** | (Number of facilities) | (Number of rooms) | (Number of facilities) | (Number of rooms) |
| Hotels/resorts | 19 | 1,684 | 119 | 5,931 |
| Tourist village | 26 | 843 | 97 | 810 |
| Homestay | | | 256 | 1,011 |
| Guest house | 34 | 482 | 6 | 70 |
| Other | | | 1 | 56 |
| *Total* | *79* | *3,009* | *479* | *7,878* |

*Source: Authors' compilation based on statistics provided by Hoi An Department of Trade and Commerce.*

300,000 tickets sold and 280,000 overnight stays in 2018. However, actual footfall seems certain to exceed 4 million if the number of tourists is multiplied by the average stay length. Prior to COVID-19, the increase of tourist arrivals had been further fueled by demand from Northeast Asia due to the rapid increase in direct flights to Da Nang international airport (Dan Tri, 2018).

Socio-economic indicators provide further evidence of the WHS's positive impact – compared to national medians, Hoi An had a higher average GDP and lower number of households living below the poverty line (UNESCO, 2008b). In 2008, Hoi An was upgraded to third-level city status and the contribution of the tourism sector to the city's GDP increased from 65% to 70%, doubling the number of jobs created from 3,411 in 2008 to 7,535 in 2016 (Table 2). The number of hotel rooms has also doubled, particularly the homestay category that added over a thousand extra rooms since its establishment in 2012. Estimated revenue from tourism has increased from 34.4 million to 1 billion USD (2007-2016), and the fees from entrance tickets from 1.7 million to 82 million (Table 2), signifying the positive impacts of tourism development on Hoi An. However, musuemification has wrought radical change (Avieli, 2015), posing serious questions for social sustainability as evidenced in the spatial and demographic transformations analyzed next.

## Museumification of the core zone

As historic towns turn into tourism destinations, landscape conservation measures rise to the fore. Moreover, Hoi An was listed just as the WHS program was seeking to shake off claims of a Eurocentric bias toward tangible heritage (Harrison & Hitchcook, 2005). Despite the prescient timing, UNESCO guidelines still tended to favor conservation of physical artifacts and thus run the risk of museumification (Logan, 2012). After Hoi An's listing, a UNESCO subsidy program was established to support the restoration of privately-owned houses, but the *de facto* prioritization of tangible heritage in the core has been criticized for turning the living town into a museum (Di Giovine, 2009a, 2009b).

The historic core's compact dimensions have also shaped the socio-spatial consequences of listing. The inscribed property of Hoi An's core covers 30 ha comprising 1,107 timber frame buildings that include architectural monuments, commercial and domestic structures (World Heritage Center Website). The town's center is too confined to accommodate much spatial distinction between the conservation and facilities zones viz-a-viz the tourist-historic city model proposed by Ashworth and Tunbridge (2000). Instead, the heritage buildings themselves have been converted into tourist facilities such as shops, coffee house or galleries while retaining the external façade of the authentic wooden shell in accordance with conservation regulations. Thus the ancient town has gradually been converted into an 'open-air museum' (Ibid, p.93) due to commercialization in the core zone. By 2008, tourism already contributed 65% of the whole town's GDP (UNESCO, 2008b) triggering a 'tremendous increase in retailing, and growth in tourism-related accommodation and services and a discernible shift in the ethos of the resident population (a growing proportion is now non-local)' (Parnwell, 2016, p.30). Consequently, the atmosphere and lifestyle of the town has changed dramatically: -

> *In the evening, the central streets [in Minh An Ward] are very crowded. The city has developed many attractions outside the centre, however, tourists from all corners of the city prefer to walk in the central area. Increased number of visitors affects the peacefulness of the city. Hoi An means 'Meeting place of Peacefulness' but now it is no longer peaceful and quiet. Hoi An is beautiful only after 9 or 10 PM after the streets become quiet (R6).*

Comments such as these suggest that the core has become overly focused on tourism at the expense of the resident population. In response, intangible heritage has been increasingly promoted as an alternative to attract more tourists and lengthen their stay. An 'adaptive re-use' policy has been advocated to, re-use, upgrade, or adjust the capacity of existing buildings to suit new conditions when the former function has been rendered obsolete (Douglas, 2006). The Hoi An Center for Monument Management and Preservation (HACMMP) has identified over 1,350 relics of heritage importance of which 1,254 are artistic or architectural structures, including shop-houses, family chapels and assembly halls (UNESCO, 2008b). The shop-houses exemplify the adaptive re-use ethos since they typically contain units of three buildings arranged together consecutively on long, narrow lots. In former times, this design 'allow[ed] several generations of a family to live close together while maintaining privacy' (Ibid, p. 15) thus helping to conserve intangible heritage value by preserving the landscape while encouraging the continuity of social life in a community (Yung & Chan, 2012). By 2009, ten years after WHS inscription, approximately 200 heritage buildings had been comprehensively restored; with 43 government-owned properties leased as residential, public and commercial spaces after restoration. The revenue generated from leased properties was used to improve public facilities and also plowed back into conservation projects. The effort of government-supported implementation of 'adaptive re-use' allowed low-income families to retain the ownership of property deeds in the face of pressure from real estate investors and domestic migrants from megacities such as Hanoi or Ho Chi Minh (Pham, 2016). Adaptive re-use thus facilitates local residents' upkeep of the old town's social fabric, encouraging social sustainability by offering an incentive to safeguard both intangible and tangible heritage. However, it has also inadvertently expedited museumification as will be discussed next.

## Demographic transformation

Mapping demographics from 2001 to 2010 within spatial conservation zones, an inverse correlation emerged between resident and tourist trends (Figure 4). Minh An Ward, which shows close territorial alignment with the WHS core zone, experienced a population decline of 20% even as the municipality's total population increased by 17% during the same period. A net flow of outward migration was also observed in Son Phong Ward, which contains both Zone I and buffer Zone IIA, and to a lesser extent in Cam Nam Ward (Zone IIA & IIB), suggesting that areas with stricter planning permission laws and more tourists are depopulating more quickly. The population increased by over 10% in Cam Chau and Cam Pho, two buffer zone wards where hotel facilities are concentrated.

The declining population in the core reflects the museumification accelerated by tourism impacts combined with stricter regulations (for example, no new hotels can be constructed in the core). Consequently it is becoming harder to verify Parnwell's (2016) claim that 'the majority of the buildings are still used by the local residents.' Nowadays

> *the residents come from different regions of the country, especially from large cities such as Hanoi, Ho Chi Minh City and Da Nang .... In the core zone area, over 80% are not original residents, but they rent or buy properties for business purposes and are eligible to register as permanent resident of the core zone (R6).*

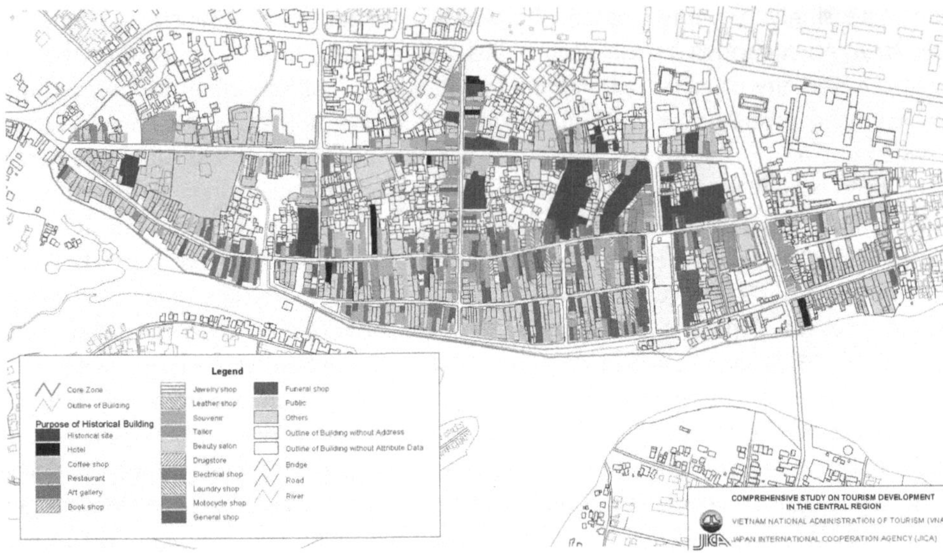


**Figure 5.** Usage purpose of historic houses in the core zone.
Source: Japan International Cooperation Agency (JICA).

Longitudinal research by Showa Women's University (Hoi An City Portal, 2017) tracked the transformation of 453 historical buildings, of which 409 are now used for retail and tourism-related services, including souvenirs, art galleries, clothing and tailoring shops. The first tailor's opened in 1995, since when Hoi An has built from scratch a reputation for haberdashery (Di Giovine, 2009b, p. 264). Together with the impact on consumption patterns, 'important dwellings have partly been 'museumised' for tourist visits, even though families continue to live in and operate their businesses from them' Parnwell (2016, p.23-4). Ando et al (2015) confirmed that few of the shops in the core still cater for local residents, and on the four main streets in Zone I, there are 38 shop-houses without any residents living on the premises. Historic buildings, especially those in prime locations along the water-front and main roads have been converted into shops, souvenir outlets, cafes and restaurants (Figure 5). Parnwell (2016) lists specific examples such as the Old House of Phung Hung (near the Japanese Bridge) and the Tran Family Home and Chapel (21 Le Loi).

The conversion of dwellings eventually pushes out existing occupants who take with them the traditional values and irreplaceable intangible heritage of local lifestyles.

> *Hoi An has become alien… Tourism brought changes to local life, making it more prosperous… But I do not feel better as I have not sighted any familiar neighbours in the entire street. I can see cracks in traditional culture and lifestyle, unauthentic behaviour and goods sold in the city. I am really worried* (R16).

The rapid rise of core zone real estate prices has triggered a range of social issues such as incoming migration and depopulation, accelerated by local government 'relocation' programs to actively thin out the population density of central Minh An (Phan, 2004). In the early 2000s, local residents were encouraged to move out of the core and purchase subsidized plots of land in the suburbs with infrastructure provided by the government. There have subsequently been multiple movements to relocate residents living in conservation zones, with some 8,000 people 'voluntarily' relocated (Hoi An City Portal, 2017).

## Revitalization of the buffer zone

Critics note that intangible heritage that is 'orchestrated to 'humanize' tourist experiences of the town – creating the sense of a 'living museum' – are of increasingly tenuous authenticity' Parnwell (2016, p.14). Together with architectural conservation, and the afore-mentioned haberdashery trade, evidence of museumification includes (re-)created customs, such as the Lantern Festival - an 'invented tradition' designed to portray an idealized vision of the past by bringing back 'a night of the old days' (Di Giovine, 2009b, p. 224). However, an alternative explanation by some respondents suggest that 'authentic' local lifestyles have simply shifted outwards spatially:-

> *The original residents sold or rent out their ancestral properties located in the core zone, eventually moving out and registering their permanent residence in the buffer zone and retaining their own lifestyle and experience (R6).*

Local authorities' efforts to capitalize on cultural activities in the buffer further support claims of social sustainability. Five craft villages located around the periphery of the buffer zone have been promoted as cultural tourism destinations, designed to extend tourists' stay and increase spending while retaining and diversifying traditional craftsmanship. Nurturing the creative spirit of a craft community can help retain traditional products, skills, local materials and artistic inspiration for local residents (Lovell & Bull, 2018). One example is Kim Bong village, located 0.5 km southwest of Hoi An's historic core, whose wood workers were renowned for their furniture, boat and pagoda construction. To revitalize this traditional craft village, wood carving workshops have been initiated and community-based tourism promoted. Building on the legacy of woodwork artisanship, the villagers have diversified to produce local souvenirs and tourist boats, and conduct workshops for both local youth and tourists (Bui & Jolliffe, 2013). Craft villages thus help transmit traditional knowledge, mitigating younger generations' reluctance to carry on their family vocation (UNESCO, 2007, p.5). Local communities have adapted to changes in physical landscapes and livelihood by uprooting and restructuring their social fabric in the buffer zone. This continuity in relocated lifestyles could be taken as evidence of the kind of social resilience suggested by Lew et al (2016). Another example is Thanh Ha Ceramic Park, a former center of earthenware household products now redeveloped as a traditional pottery. Yet despite promotional efforts, tours to remote Kim Bong and Thanh Ha remain significantly less popular than to the more recently established Silk Village, conveniently located on the main road from Da Nang. Spatial dispersal of tourists away from the congested core toward the craft villages in the buffer zone thus remains an unfulfilled mitigation strategy. For example, only five percent of all Hoi An visitors surveyed by ESRT (2014) joined optional tours to the aforementioned craft villages in the buffer zone. Moreover, question marks remain over the authenticity of visitors' experience (McKercher & du Cros, 2003). For example, another effort to revitalize Cam Nam island in the buffer zone is a new cultural show entitled 'Hoi

An Memories' that features a theme park with choreographed dance, music performance and light display.

## Discussion

Drawing on multiple methods and data sources, this research investigated the long-term transformation of zoning for conservation against a backdrop of tourism development and demographic change in a Southeast Asian WHS. Core and buffer zones feature so universally in management plans of both natural and cultural WHSs that they are considered to be neo-normative (Gillespie, 2012). Yet longitudinal analyses of the socio-spatial consequences of such zoning remain sparse. Our findings from Hoi An Ancient Town show: an inverse correlation between trends in tourist numbers and resident populations in the core and buffer zones; a spatial zoning hierarchy that struggles to accommodate the socio-economic fabric of a living WHS; and the need for a reappraisal of the conventional Eurocentric approach to zoning WHSs.

Two decades of rapid tourism growth has expedited a process of museumification within Hoi An's core, and triggered reflexive efforts to revitalize the buffer zone. Although the core zone's on-going museumification represents a gentrification variant, a clear distinction can be made from Gotham's (2005) definition based largely on evidence from Western European and North America. That model was characterized by gradual displacement (Hackworth, 2002) of newly-wealthy residents moving 'back to the city center' in a reversal of suburbanization (Smith, 2002; Butler & Lees, 2006). Conversely, the migration trends at Hoi An appear closer to Chan's (2016) 'self-led' and 'state-led' gentrification. While locals are displaced from the core, the buffer zone is a spillover space where customs and remnant life styles converge. Tourist demand for authentic experiences could bring opportunities for 'self-gentrification' in small scale accommodation facilities or conversion of residence places into shops for tourists. However, in this era of rapid change, certain intangible heritage such as everyday lifestyles may struggle to grow new roots after transplanting to the buffer zone and could be jettisoned in place of tourism development. Local residents can feel alienated by transformation to meet tourists' needs. Growing sentimental detachment from a place on the behalf of local residents is reinforced by price rises in real estate, rent and services. Conversion of well-preserved or renovated residences into upmarket bars and restaurants resonate with the findings of Teo and Huang (1995) in Singapore. Meanwhile, an influx of migrant laborers for the cafes, stalls and haberdasheries echoes other WHSs such as Lijiang where new souvenir shops are 'overwhelmingly controlled or operated by people from the coastal areas of China (namely domestic migrants from Gunagdong and Fujian) and from overseas (e.g. South Korea and Japan)' (Su & Teo, 2008, p.157).

Su, Wall and Xu (2016) also discussed short-term economic benefits of WHSs at the expense of long-term social sustainability due to displacement of locals from the core. Physical eviction without compensation has not occurred at Hoi An, yet our findings still reiterate that the long-term consequences of WHS status are complicated, with residents displaced incrementally via the push and pull forces of zoning and tourism demand. Newly-listed or tentative WHSs should be monitored closely to avoid bipolar development and prevent extreme examples of museumification such as Borobudur in Indonesia, Sukhothai in Thailand, and Petra in Jordan (Poulios 2011; Nagaoka, 2016).

Evidence from Hoi An supports previous research on the magnetic attraction of WHS listed status and its 'tourism enhancing effect' (Yang, Lin & Han, 2010), exemplified by the contribution of tourism to GDP growth, job creation and poverty alleviation. However, socio-economic consequences of listing should not be overlooked. These include an increase in land prices and changing demography driving locals out of the core together with their 'traditional lifestyle, religion, customs and cooking' (Avieli, 2015). The case of Hoi An also illustrates the combination of conventional and novel approaches in heritage conservation planning: the material fabric-based approach, reflected in government initiatives to purchase endangered old houses and invest in those properties' restoration; and the living approach reflected in the UNESCO program of 'Adaptive Re-use of Built Heritage' (Pham, 2016, p.279). Erosion of local lifestyles and customs is a consequence of a Euro-centric planning norm that prioritizes physical artifacts of 'high-culture' heritage over the social fabric of everyday lifestyles. However, 'adaptive re-use' can also lead to gentrification that compromises the authenticity of an area and the continuity of the residents' lifestyles (Urry, 2002).

Findings from this research thus question the effectiveness of current zoning approaches for conservation of a living heritage site. Hoi An's authentically preserved architecture is exceptional by national, or even international, standards. Yet human interactions with heritage are complicated and the socio-spatial zones appear to have inadvertently corroborated King's (2016) observation that 'the living cultures which energize a site are often marginalized or traditionalised'. The Hoi An case demonstrates that UNESCO has acknowledged certain limitations of the buffer zone's effectiveness but is unwilling or unable to instigate meaningful change (Avieli, 2015). Clustering of funding and conservation efforts in the core results in growing disparities with the buffer zone that becomes a repository for the displaced local residents whose lifestyles represent a vital spark for intangible heritage. Although the city government is actively promoting craft villages around the buffer zone's perimeter, much work remains to maximize tourism's potential to contribute to social sustainability by curbing bipolar development.

Finally, the notion of continuity of Asian living heritage evidenced in our research demands urgent reappraisal of UNESCO approaches to WHS selection and post-inscription management. The WHS program's underlying philosophy of 'outstanding universal value' has itself been criticized as a top-down, Eurocentric approach (Winter, 2014) that does not always coincide with host communities' own idea of the significance of heritage. In order for local communities to feel a continuing sense of ownership and involvement, further reconciliation is required with UNESCO planning guidelines. Yet emerging economies face particular barriers to WHS management, including the community role in decision making, benefit-sharing, ownership of historic places and artifacts, and forced displacement (Timothy & Boyd, 2006). Reducing the Euro-centric character of the WHS program requires broader heritage discourse, and acknowledging variation in global concept and practice (Logal, Kockel, & Craith, 2016). In a Southeast Asian context, heritage appreciation is said to stem from its cultural 'pluralism' and 'continuity' (Chapagain, 2013), so this longitudinal research of Hoi An's spatial and social transformation could provide insights into the conservation of other living WHSs, such as Georgetown, Lijiang, and Luang Prabang. Unlike in

Europe, where a heritage industry has emerged gradually supported by established legislation, policy and control mechanisms, in Southeast Asian LDCs, development pressure remains greater and heritage funds fewer. Conservation models cannot simply replicate the European equivalents (Orbasli, 2000) but must innovate to build better site-specific philosophies into planning and development, rethinking visitor management tools that are outdated, inadequate or struggling to keep pace with tourism growth (Hitchcock et al., 2010). Many Southeast Asian WHSs from Angkor to Lijiang suffer from 'over-tourism' and Hoi An now faces familiar issues coping with congestion in the core even as buffer zone attractions lie latent or underutilized. Our study bridges a missing link between social sustainability and spatial planning. Based on evidence of the demographic transition of tourists and residents in the core and buffer zones, social sustainability centered on adaptative re-use is a complex process that belies simple mitigation efforts. Ultimately, Hoi An has witnessed an interlinked intensification of tradeoffs between living and built heritage; adaptive re-use and museumification; and social sustainability of residents reshaped by spatial planning regulations. As the 'over-tourism' debate deepens, 'concern with gentrification of historic cities has been almost non-existent in the ICOMOS valuation discourse' to date (Vahtikari, 2019: p. 113), but these results from Hoi An suggest the zoning debate to be of paramount importance.

## Conclusion

In the contested terrain between tourism and migration (Williams & Hall, 2000), considerable scope remains for assessing the broader social sustainability implications of listing as a UNESCO World Heritage Site (WHS). This longitudinal case study of Hoi An, an archetypal Southeast Asian WHS 'success-story,' probes the interlocking mechanisms of museumification and gentrification that result from the WHS program's twin goals of conservation and tourism development. Focusing on the socio-spatial consequences of zoning, a neo-normative UNESCO planning principle, we examine social sustainability in the core and buffer zones from the dual perspectives of development and demography. The mixed-method approach triangulated interviews with secondary data from government statistics, reports, maps and planning guidelines.

Findings show an exponential increase in visitor numbers and GDP growth since Hoi An was listed as a WHS in 1999. Tourism development has helped lift a generation of Hoianese out of poverty while generating considerable funds for heritage conservation. Beyond the direct benefits due to WHS-related investment, subsidies and revenue from entrance tickets to Hoi An's historical center, indirect economic impacts via the multiplier effect have facilitated authentic architectural renovation of the core zone. However, negative impacts of WHS status entail consequences both economic (e.g. rising real estate prices) and social (e.g. stricter planning regulations and more tourist congestion). Such impacts have driven depopulation of the core, evidenced by a 20% decline in residents of central Minh An Ward during a period from 2000 to 2010 even as the overall population across Hoi An municipality increased by 17%.

In sum, the hierarchical WHS zoning rationale has legitimized bipolar development of Hoi An's buffer zone in tandem with 'museumification' of the core. The 'adaptive re-use' of traditional wooden buildings has facilitated the conversion of private family residences into shops, restaurants or other commercial service providers. These findings further the debate over the long-term social sustainability of WHSs by underlining the need for more empirical investigation of the spatial consequences of core and buffer zones. As Hoi An's core transformed from living heritage into a tourism stage, demographic and development trends in the buffer zone reinforced its subordinate role in a self-fulfilling cycle. Aggressive museumification of the historic center, together with unfettered development of the buffer zone thus sends warning signs to WHS planners and site managers, although a positive counter-strategy is also discernible at Hoi An from ongoing attempts to expand and strengthen the buffer zone via a network of craft villages dotted around the perimeter.

Lastly, a number of limitations in this paper are acknowledged. First, the single-case study method raises questions of representativeness (Yin, 2009). To take a broader view and improve transferability, a series of case studies and cross-cultural comparisons would allow more parallels to be drawn with the perennial planning dilemma of zoning at other Southeast Asian WHSs such as Lijiang (China); Angkor Watt (Cambodia); and Luang Prabang (Laos PDR). A quantitative survey of local residents and tourists in different zones could also uncover measureable perception gaps in the socio-spatial consequences of the core-and-buffer zones. Nonetheless, the paper provides empirical insights into WHS management, and a long-term Southeast Asian perspective on the ongoing tourism gentrification debate to counter models from the US and China (Gotham, 2005; Liang & Bao, 2015).

## Disclosure statement

No potential conflict of interest was reported by the author(s).

## References

Adie, B. A., Hall, C. M., & Prayag, G. (2018). World Heritage as a placebo brand: A comparative analysis of three sites and marketing implications. *Journal of Sustainable Tourism*, *26*(3), 399–415. https://doi.org/10.1080/09669582.2017.1359277

Ando, K., Fukukawa, Y., & Tomoda, H. (2015). The formulation process and characteristics of historic conservation system in Hoi An, Vietnam. *Journal of Architecture and Planning (Transactions of Aij)*, *80*(708), 379–389 (in Japanese). https://doi.org/10.3130/aija.80.379

Ashworth, G. J., & Tunbridge, J. E. (2000). *The tourist – historic cities*. Elsevier. https://www.sciencedirect.com/book/9780080436753/the-tourist-historic-city#book-description

Avieli, N. (2015). The rise and fall (?) of Hoi An, a UNESCO World Heritage Site in Vietnam. *Journal of Social Issues in Southeast Asia*, *30*(1), 35–71. Project MUSE muse.jhu.edu/article/577529 https://doi.org/10.1355/sj30-1b

Badcock, B. (2001). Thirty years on: Gentrification and class changeover in Adelaide's inner suburbs, 1966-96. *Urban Studies*, *38*(9), 1559–1572. https://doi.org/10.1080/00420980120080441

Brooks, G. (2008). Exploiting the benefit of World Heritage Listing: Evora, Portugal and Hoi An, Vietnam. In B. Hayllar, T. Griffin, & D. Edwards (Eds.), *City spaces–tourist places* (pp. 341–356). Elsevier.

Bui, H. T., & Jolliffe, L. (2013). Souvenir production and attraction: Vietnam Traditional Craft Village. In J. Cave, L. Jolliffe, & T. Baum (Eds.), *Tourism and souvenir: Global perspectives from the margins* (pp 161–175). Channel View Publications.

Bui, H. T., Jones, T. E., Weaver, D. B., & Le, A. (2020). The adaptive resilience of living cultural heritage in a tourism destination. *Journal of Sustainable Tourism*, *28*(7), 1022–1040. https://doi.org/10.1080/09669582.2020.1717503

Bui, H. T., & Le, T. A. (2018). Case study Vietnam: Climate change impacts on UNESCO World Heritage – the case of Hoi An ancient town. In A. Jones & M. Phillips (Eds.), *Global climate change and coastal tourism* (pp. 191–199). CABI.

Butler, T., & Lees, L. (2006). Super-gentrification in Barnsbury, London: Globalization and gentrifying global elites at the neighbourhood level. *Transactions of the Institute of British Geographers*, *31*(4), 467–487. https://doi.org/10.1111/j.1475-5661.2006.00220.x

Chan, J. H., Iankova, K., Zhang, Y., McDonald, T., & Qi, X. (2016). The role of self-gentrification in sustainable tourism: Indigenous entrepreneurship at Honghe Hani Rice Terraces World Heritage Site, China. *Journal of Sustainable Tourism*, *24*(8-9), 1262–1279. https://doi.org/10.1080/09669582.2016.1189923

Chapagain, N. (2013). Introduction: Contexts and concerns in Asian heritage management. In K. Silva & N. Chapagain (Eds.), *Asian heritage management: Context, concerns and prospects* (pp.1–30). Routledge.

Davidson, M., & Lees, L. (2005). New-build 'gentrification' and London's riverside renaissance. *Environment and Planning A: Economy and Space*, *37*(7), 1165–1190. https://doi.org/10.1068/a3739

Davidson, M., & Lees, L. (2009). New-build gentrification: Its histories, trajectories, and critical geographies. *Population, Space and Place*, *16*(5), 411. https://doi.org/10.1002/psp.584

Dempsey, N., Bramley, G., Power, S., & Brown, C. (2011). The social dimension of sustainable development: Defining urban social sustainability. *Sustainable Development*, *19*(5), 289–300. https://doi.org/10.1002/sd.417

Di Giovine, M. A. (2009a). *The heritage-scape: UNESCO, World Heritage and tourism*. Lexington Books.

Di Giovine, M. A. (2009b). Revitalization and counter-revitalization: Tourism, heritage, and the Lantern Festival as catalysts for regeneration in Hoi An, Vietnam. *Journal of Policy Research in Tourism, Leisure and Events*, *1*(3), 208–230. https://doi.org/10.1080/19407960903204364

Douglas, J. (2006). *Building adaptation* (2nd ed.). Butterworth-Heinemann.

ESRT (2014). *Visitor survey*. European Union funded Environmentally & Socially Responsible Tourism (ESRT) Capacity Development Programme.

Fairclough, N. (2003). *Analyzing discourse: Textual analysis for social research*. Routledge.

Gillespie, J. (2012). Buffering for conservation at Angkor: questioning the spatial regulation of a World Heritage property. *International Journal of Heritage Studies*, *18*(2), 194–208. https://doi.org/10.1080/13527258.2011.603737

Gotham, K. F. (2005). Tourism gentrification: The case of New Orleans' Vieux Carre (French Quarter). *Urban Studies*, *42*(7), 1099–1121. https://doi.org/10.1080/00420980500120881

Guest, G., MacQueen, K., & Namey, E. (2012). *Applied Thematic Analysis*. Sage.
Hackworth, J. (2002). Postrecession gentrification in New York City. *Urban Affairs Review, 37*(6), 815–843. https://doi.org/10.1177/107874037006003
Harrison, D., & Hitchcook, M. (2005). *The politics of World Heritage: Negotiating tourism and conservation*. Channel View Publications.
Hassan, M., & Xie, H. (2020). *Climate change and conservation of coastal built heritage*. Springer. https://link.springer.com/book/10.1007/978-981-13-8672-5
He, S. (2009). New-build gentrification in Central Shanghai: Demographic changes and socioeconomic implications. *Population, Space and Place, 16*(5), 361. https://doi.org/10.1002/psp.548
Hines, J. D. (2012). The post-industrial regime of production/consumption and the rural gentrification of the New West Archipelago. *Antipode, 44*(1), 74–97. https://doi.org/10.1111/j.1467-8330.2011.00843.x
Hitchcock, M., King, V. T., & Parnwell, M. (2010). *Tourism in Southeast Asia: challenges and new directions*. NIAS Press.
Hoi An City Portal (2017). Retrieved from http://hoian.gov.vn/default.aspx on Mar. 14, 2018.
Hoyle, B. S. (1988). *Revitalising the waterfront: International dimensions of dockland redevelopment*. Belhaven.
Hubbard, P. (2008). Regulating the social impacts of studentification: A Loughborough case study. *Environment and Planning A: Economy and Space, 40*(2), 323–341. https://doi.org/10.1068/a396
Jimura, T. (2011). The impact of world heritage site inscription on local communities: A case study of Ogimachi, Shirakawa-mura, Japan. *Tourism Management, 32*(2), 288–296. https://doi.org/10.1016/j.tourman.2010.02.005
King, V. (Ed.). (2016). *UNESCO in Southeast Asia: World Heritage Sites in comparative context*. NIAS Press.
Lew, A. A., Ng, P. T., Ni, C. C., & Wu, T. C. (2016). Community sustainability and resilience: Similarities, differences and indicators. *Tourism Geographies, 18*(1), 18–27. https://doi.org/10.1080/14616688.2015.1122664
Liang, Z.-X., & Bao, J. G. (2015). Tourism gentrification in Shenzhen, China: causes and socio-spatial consequences. *Tourism Geographies, 17*(3), 461–481. https://doi.org/10.1080/14616688.2014.1000954
Liang, Z. X., & Hui, T. K. (2016). Residents' quality of life and attitudes toward tourism development in China. *Tourism Management, 57*, 56–67. https://doi.org/10.1016/j.tourman.2016.05.001
Logal, Kockel, & Craith. (2016). The new heritage studies: origins and evolution, problems and prospects. In W. Logan, N. Craith, & U. Kockel (Eds). *A companion to heritage studies* (pp.1–25). Wiley.
Logan, W. (2002). *The disappearing 'Asian' city: Protecting Asia's urban heritage in a globalising world*. Oxford University Press.
Logan, W. (2012). States, Governance and the Politics of Culture: World Heritage in Asia. In P. Daly & T. Winter (Eds) *Routledge handbook of heritage in Asia*. (pp. 113–128). Routledge.
Loulanski, T., & Loulanski, V. (2011). The sustainable integration of cultural heritage and tourism: a meta-study. *Journal of Sustainable Tourism, 19*(7), 837–862. https://doi.org/10.1080/09669582.2011.553286
Lovell, J., & Bull, C. (2018). *Authentic and inauthentic places in tourism: From heritage sites to theme park*. Routledge.
Lowenthal, D. (2015). *The past is a foreign country - revisited*. Cambridge University Press.
McKercher, B., & Du Cros, H. (2003). Testing a cultural tourism typology. *International Journal of Tourism Research, 5*(1), 45–58. https://doi.org/10.1002/jtr.417
Miura, K. (2011). World Heritage Making in Angkor. Global, Regional, National and Local Actors, Interplays and Implications. In B. Hauser-Schäublin (Edt) *World Heritage Angkor and beyond circumstances and implications of UNESCO listings in Cambodia*. (pp. 9–33). Universitätsverlag.
Nagaoka, M. (2016). *Cultural landscape management at Borobudur, Indonesia*. Springer. URL: https://link.springer.com/book/10.1007/978-3-319-42046-2
Orbasli, A. (2000). *Tourists in historic towns*. Taylor & Francis.

Orbasli, A., & Woodward, S. (2009). Tourism and heritage conservation. In T. Jamal & M. Robinson (Eds.) *The Sage handbook of tourism studies*. (pp. 314–332). Sage Publications.

Parnwell, M. (2016). Heritage management and tourism development in Hoi An, Vietnam. In V. King (Ed.), *UNESCO in Southeast Asia: World Heritage Sites in comparative context*. NIAS Press.

Pham, H. (2016). Living heritage, community participation and sustainability – Redefining development strategies sin the Hoi An Ancient Town World Heritage Property, Vietnam. In S. Labadi & L. William (Eds) *Urban heritage, development and sustainability*. (pp. 274–290). Routledge.

Phillips, M. (2004). Other geographies of gentrification. *Progress in Human Geography*, *28*(1), 5–30. https://doi.org/10.1191/0309132504ph458oa

Poulios, I. (2011). Is Every Heritage Site a 'Living' One? Linking Conservation to Communities' Association with Sites. *The Historic Environment: Policy & Practice*, *2*(2), 144–156. https://doi.org/10.1179/175675011X13122044136550

Poulios, I. (2014). Discussing strategy in heritage conservation: living heritage approach as an example of strategic innovation. *Journal of Cultural Heritage Management and Sustainable Development*, *4*(1), 16–34. https://doi.org/10.1108/JCHMSD-10-2012-004 https://doi.org/10.1108/JCHMSD-10-2012-0048

Prime Minister's Office (2012). Approval of General Planning for Conservation, Reconstruction and Development of UNESCO World Cultural Heritage – Hoi An Ancient Town For the Development of Hoi An City and Tourism period 2012 – 2025 – Decision number: 78/QĐ-TTg. Hanoi: The Prime Minister's Office.

Rasoolimanesh, S. M., Jaafar, M., Kock, N., & Ahmad, A. G. (2017). The effects of community factors on residents' perceptions toward World Heritage Site inscription and sustainable tourism development. *Journal of Sustainable Tourism*, *25*(2), 198–216. https://doi.org/10.1080/09669582.2016.1195836

Reeves, K., & Long, C. (2011). Unbearable pressures on paradise? Tourism and heritage management in Luang Prabang, a World Heritage site. *Critical Asian Studies*, *43*(1), 3–22. https://doi.org/10.1080/14672715.2011.537849

Shin, H. B. (2010). Urban conservation and revalorisation of dilapidated historic quarters: The case of Nanluoguxiang in Beijing. *Cities*, *27*(1), S43–S54. https://doi.org/10.1016/j.cities.2010.03.006

Smith, N. (2002). New globalism, new urbanism: Gentrification as global urban strategy. *Antipode*, *34*(3), 427–450. https://doi.org/10.1111/1467-8330.00249

Smith, D. P., & Holt, L. (2007). Studentification and 'apprentice' gentrifiers within Britain's provincial towns and cities: Extending the meaning of gentrification. *Environment and Planning A: Economy and Space*, *39*(1), 142–161. https://doi.org/10.1068/a38476

Su, X., & Teo, P. (2008). Tourism politics in Lijiang, China: An analysis of state and local interactions in tourism development. *Tourism Geographies*, *10*(2), 150–168. https://doi.org/10.1080/14616680801999992

Su, M. M., & Wall, G. (2012). Global–local relationships and governance issues at the Great Wall World Heritage Site, China. *Journal of Sustainable Tourism*, *20*(8), 1067–1086. https://doi.org/10.1080/09669582.2012.671330

Su, M. M., Wall, G., & Xu, K. (2016). Heritage tourism and livelihood sustainability of a resettled rural community: Mount Sanqingshan World Heritage Site, China. *Journal of Sustainable Tourism*, *24*(5), 735–757. https://doi.org/10.1080/09669582.2015.1085868

Teo, P., & Huang, S. (1995). Tourism and heritage conservation in Singapore. *Annals of Tourism Research*, *22*(3), 589–615. https://doi.org/10.1016/0160-7383(95)00003-O https://doi.org/10.1016/0160-7383(95)00003-O

Timothy, D. J., & Boyd, S. W. (2006). Heritage tourism in the 21st century: Valued traditions and new perspectives. *Journal of Heritage Tourism*, *1*(1), 1–16. https://doi.org/10.1080/17438730608668462

Tri, D. (2018). International tourist trends, an interview for Dan Tri Newspaper with Hoi An Department of Trade and Commerce. https://dulich.dantri.com.vn/du-lich/hoi-an-luong-khach-tang-nhung-khach-tu-chau-au-bac-my-giam-manh-20181105081329581.html. on 13.7.2019

Trinh, T. T., Ryan, C., & Cave, J. (2014). Souvenir sellers and perceptions of authenticity–The retailers of Hội An, Vietnam. *Tourism Management*, *45*, 275–283. https://doi.org/10.1016/j.tourman.2014.05.005

UNESCO (2007). Asia conserved: Lessons learned from the UNESCO Asia-Pacific
UNESCO (2008a). *The heritage homeowner's preservation manuals for UNESCO World Heritage Sites*. UNESCO. URL:https://unesdoc.unesco.org/ark:/48223/pf0000193014
UNESCO (2008b). *Cultural tourism and heritage management in the World Heritage Site of Ancient Town of Hoi An, Vietnam*. UNESCO. https://unesdoc.unesco.org/ark:/48223/pf0000182646
UNESCO (2008c). *International expert meeting on World Heritage and buffer zones*. https://whc.unesco.org/en/events/473/ Retrieved on May, 5, 2020.
UNESCO (2017). *The Hoi An declaration on urban heritage conservation and development in Asia*. UNESCO. URL:http://www.unesco.org/new/en/hanoi/about-the-ha-noi-office/single-view/news/hoi_an_hosts_international_conference_on_urban_heritage_cons/.
Urry, J. (2002). *The tourist gaze*. (2nd ed.). Sage.
Vahtikari, T. (2019). *Valuing World Heritage cities* (1st ed.). Routledge.
Wescott, G., & Molinski, J. (1993). Loving our parks to death. *Habitat Australia*, *21*(1), 12–19.
Williams, A. M., & Hall, C. M. (2000). Tourism and migration: New relationships between production and consumption. *Tourism Geographies*, *2*(1), 5–27. https://doi.org/10.1080/146166800363420
Winter, T. (2014). Beyond Eurocentrism? Heritage conservation and the politics of difference. *International Journal of Heritage Studies*, *20*(2), 123–137. https://doi.org/10.1080/13527258.2012.736403
Winter, T., & Daly, P. (2012). Heritage in Asia: Converging forces, conflicting values. In P. Daly & T. Winter (Ed.), *Routledge handbook of heritage in Asia* (pp. 1–35). Routledge.
Xu, H., Zhu, D., & Bao, J. (2016). Sustainability and nature-based mass tourism: lessons from China's approach to the Huangshan Scenic Park. *Journal of Sustainable Tourism*, *24*(2), 182–202. https://doi.org/10.1080/09669582.2015.1071381
Yang, C.-H., Lin, H.-L., & Han, C.-C. (2010). Analysis of international tourist arrivals in China: The role of World Heritage sites. *Tourism Management*, *31*(6), 827–837. https://doi.org/10.1016/j.tourman.2009.08.008
Yang, Y., Xue, L., & Jones, T. E. (2019). Tourism-enhancing effect of World Heritage Sites: Panacea or placebo? A meta-analysis. *Annals of Tourism Research*, *75*, 29–41. https://doi.org/10.1016/j.annals.2018.12.007
Yin, R. K. (2009). *Case study research: Design and method* (5th ed.). Sage Publications.
Yung, E. H. K., & Chan, E. H. W. (2012). Implementation challenges to the adaptive reuse of heritage buildings: Towards the goals of sustainable, low carbon cities. *Habitat International*, *36*(3), 352–361. https://doi.org/10.1016/j.habitatint.2011.11.001
Zerva, K., Palou, S., Blasco, D., & Benito, J. A. D. (2019). Tourism-philia versus tourism-phobia: residents and destination management organization's publicly expressed tourism perceptions in Barcelona. *Tourism Geographies*, *21*(2), 306–329. https://doi.org/10.1080/14616688.2018.1522510

# Perceptions of and interactions between locals, migrants, and tourists in South Tyrol

Anja Marcher, Ingrid Kofler, Elisa Innerhofer and Harald Pechlaner

**ABSTRACT**

The nexus of migration and tourism is well explored, and both phenomena possess push- and pull factors. Their influence and connections can be seen in one of the top tourist destinations in Europe, South Tyrol, a region located in Northern Italy, and one of its most important tourism hubs: Ortisei. The area is particularly interesting for researchers due to its long history of tourism and the multilingualism of the region. Our in-depth and exploratory research approach helps gain a better understanding of the complexity of the migration-tourism nexus from a local perspective. It confirms the attractiveness of the tourism sector for migrants, but also the impact of both phenomena on the region. The findings show that the presence of migrants in a destination is strongly connected to the inflow of (domestic and international) tourists and the touristic experiences offered at the local level. Therefore, tourism contributes to the generation of in-migration. The growing share of migrants working in the local tourist industry represents the most evident link between the two phenomena, supporting the tourism-led migration hypothesis. But the local perception of tourists and migrants as the 'other' and trust as a relational characteristic joins the two phenomena on a different level, as both are seen as 'non-local' groups. However, within a tourist destination, both migrants and tourists interrelate with the local community, thus shaping it. And even though the statistical categories prevail based on the length of or reason to stay, at second glance those boundaries blur, and the acceptance of 'new' populations groups living, working and visiting the destination seems to increase with their duration of stay.

**摘要**

迁移和旅游的关系已经得到了很好的探讨，这两种现象都具有推动和拉动因素。它们的影响力和联系可以在欧洲最著名的旅游目的地之一，意大利北部的南提洛尔地区，以及意大利最重要的旅游枢纽之一:奥蒂塞伊看到。由于该地区有着悠久的旅游发展的历史和多种语言的使用，研究人员对该地区特别感兴趣。我们详细的探索性研究方法有助于从地方角度更好地理解迁移-旅游关系的复杂性。它证实了旅游业对移民的吸引力，但也证实了这两种现象对该地区的影响。调查结果表明，移民在目的地的出现与国内和国际游客的流入和地方层面提供的旅游体验密切相关。因此，旅游业促进了移民的产生。在当地旅游业工作的移民人数越来越多，这是这两种现象之间最明显的联系，支持了以旅游业为主导的移民假说。但是当地人把游客和移民认识为"他者"，而信任则是一种关系特征，在不同的层面上将这两种现象结合在一起，因为两者都被视为"非本地"群体。然而，在旅游目的地内，移民和游客都与当地社区相互联系，从而塑造了当地社区。尽管统计上的分类是根据逗留的时间或理由来划分的，但再看一看，这些界限就变得模糊了，人们对居住、工作和访问目的地的"新"人口群体的接受程度似乎是随着逗留时间的延长而增加的。

## Introduction

In recent decades, there has been an increased research interest in the connections, overlaps and contradictions of tourism and migration (Holert & Terkessidis, 2006; Williams & Hall, 2000). Due to advancing globalization, the complexity of different forms of population movements and trends is increasing.

To understand the nexus of migration and tourism in a destination, we examined the relationship between tourism and migration numbers at a municipal level and the perceptions of people living and working in Ortisei, a small village located in the intensive and highly developed tourist region South Tyrol in Northern Italy. We first analyzed the number of foreigners (migrants) in relation to various tourism indicators, and then conducted a more in-depth-analysis of the perceptions of and interactions between migration and tourism in the destination. The aim was to explore the link between the number of migrants living in a place and high numbers of tourists or the tourism infrastructure, and to show that, within a tourist destination, both migrants and tourists interrelate with the local community, thus shaping it. We investigate in particular how and why a relationship exists between tourism and migration, putting a special focus on the tourism-led-migration hypothesis (Williams & Hall, 2000). The relationship between the number of migrants and tourism – addressing both tourists and tourist establishments – is of particular interest. Furthermore, we wanted to understand how the presence of tourists and migrants in a destination shape the worldview of the inhabitants. After a brief introduction to theoretical issues, the methods and analytical strategy are presented, followed by a detailed description of our case study and its results, focusing on the research questions 'What is the connection between migration and tourism at the local and regional level of a destination?', 'How do migration and tourism impact the local level and how are they interrelated?'and 'What kind of mobile actors can be identified at a destination level from a local perspective?'

## Theoretical background

Along with migration (or the term migrant), travel and tourism are terms used to describe movements of people (tourists, travelers, visitors). But travel and tourism represent also an economic sector and/or a broad system of interacting relationships between people, where technical definitions to identify a specific population related to the phenomenon are distinguishable through travel purposes and its temporary, voluntary character (Hall, 2005). To define migration and tourism, concepts of space and time are essential (Hall, 2005; Glasze & Pott, 2014), but the boundaries between different definitions are fluid and often hard to distinguish. Behind the static definitions of who is a 'migrant' or a 'tourist', their mobility is a common element. If we look at tourism and migration as two 'principal elements of human mobility' (Illés & Michalkó, 2008, p. 98), both types of mobile actors could be interpreted according to Rode as follows: '[T]ourists are usually seen as the best form of human mobility because it is short term (up to one year) and mostly based on investment and consumption, while migration is viewed mostly as problematic because it introduces politic-administrative problems of citizenship, human rights, racism, and integration' (2017, p. 3–4; see also Koslowski, 2018). This becomes more evident when it comes to spatial overlapping or spaces of encounter of different types of (im)mobile people, i.e., tourists, migrants, and/or locals, as shown in a variety of research studies regarding the Mediterranean Sea, where refugees and tourists meet in the same places (Friese, 2014; Kopnina, 2007; Lenz, 2010). It underlines the notion that spatial movements also entail social mobility and are connected to status and power (Kalir, 2013).

Despite these overlaps and difficulties differentiating between tourism and migration, Williams and Hall (2000) identified two general symbiotic relationships between the two phenomena, where one generates the other. First, migrations that generates tourism flows through personal relationships such as friendship and kinship. Second, tourism that leads to migration. Migration flows are generated through the demand for labor of the local tourism industry, which can be caused by economic divergences between countries. For example, the shortage and increased demand for workforce in certain destinations and limited work opportunities or low incomes in countries of origin impact the migration decisions and individual destination preferences. These symbiotic relationships of both phenomena also represent evidence of how transnational spaces can be created through the migration-tourism nexus and are maintained and created through people connected to different places. Following that idea of the interrelation between the two phenomena, a large number of studies analyze aggregated data at a national level in order to understand the causalities between migration and tourism (Etzo, Massidda, & Piras, 2014; Seetaram & Dwyer, 2009).

Studies on global flows of humans, but also of non-humans and objects, became a central analytical approach to gain a different perspective of travel and migration through the analytical lens of mobility – or immobility (Glick Schiller & Salazar, 2013; Sheller & Urry, 2016a, 2016b). Even though people have long been mobile with fluid, multiple, and contextualized identities (Salazar, 2018, p. 153), this approach is new in the way it looks at these aspects and offers a different way to think about society (Glick Schiller & Salazar, 2013). Therefore, the way of looking at different types of travelers, such as tourists, students, refugees, labor migrants, returnees, and other 'movers', and the implications of those movements has changed (Sheller & Urry, 2016a, 2016b; Urry, 2007). The variety of types of people who are characterized by aspects of mobility (or immobility) alone illustrates how difficult it is to find or establish an overall theory. However, the mobility perspective offers a new way to look at the world, to understand global dynamics and their local nexus, and to see transnational aspects differently.

Considering these reflections on and complexities of (global) trends, challenges, and opportunities regarding migration and tourism, we aim to contribute to a better understanding of the migration-tourism nexus by answering the following research questions: What is the connection between migration and tourism at the local and regional level of a destination? How do migration and tourism impact the local level and how are they interrelated? What kind of mobile actors can be identified at a destination level from a local perspective? Starting here and with the aim of identifying the different relationships between migration and tourism, we investigated the relationship between tourism and migration at an aggregated level, using classical statistical (static) data of tourism. In a second step, we explored the nexus with a qualitative, inductive research approach by focusing on the local reality within the selected tourist destination.

## Method

To explore the nexus of migration and tourism, we applied two different methods at different spatial scales: first, we analyzed the relationship between variables measuring migration and tourism at a regional level of the Autonomous Province of Bolzano, hereinafter referred to as South Tyrol, using statistical data at the municipal level, exploring potential links between a high number of migrants and its tourist attractiveness. In the next step, to get new and additional insights, we applied an explorative and qualitative research approach focusing on one selected municipality with a long history as a tourism destination.

### *Quantitative research strategy*

Secondary data from 2017 was analyzed at the provincial level, including data of all 116 municipalities of South Tyrol (Provincial Statistics Institute [Astat], 2018a, 2018b). We focused on small local administrative units rather than focusing on the area as a whole, viz. the nation-state. The presumed causal relationship between the numbers of migrants and a variety of tourism indicators was tested via correlation using the statistical software SPSS. Migration data is illustrated by the number of non-Italian citizens living in each municipality. Tourism indicators included the number of accommodation facilities; accommodation density (number of accommodations in proportion to the municipal area); the number of beds; the bed occupation rate; overnight stays; arrivals per year (domestic and international); the number of tourism enterprises; the number of persons employed in tourism (%); the number of employed persons with an Italian or EU-15 citizenship, from new EU member states or with no EU-28-citizenship. However, not all the variables possess a normal distribution, such as the bed occupation rate. Therefore, we conducted bivariate correlations using both Pearson and Spearman correlation coefficients. Outliers were detected based on values of distribution, boxplots and normality plots with tests as they may bias the correlation. The provincial capital Bolzano and the city of Merano were excluded as outliers.

### *Qualitative research strategy*

To get a more in-depth analysis of the two phenomena and their interlinking factors, we investigated the social structure of a specific municipality, applying an explorative research approach. We conducted six in-depth interviews with various key informants, (experts living and working in the municipality), focusing on the current roles of tourism and migration in Ortisei as well as their impact on the daily life and cohabitation of its citizens. Three interview partners were representatives of the tourism sector: the director of the local destination management organization, an entrepreneur and an employee with a migration background working in the tourism sector in Ortisei for many years who had contact with many other (migrant and non-migrant) workers. The other informants were part of the local municipal administration and gave insights into economic and social developments and the current situation of the community: the local mayor, the person responsible for social affairs and the person responsible for integration. In order to identify key informants, we followed a snowball sample generation, starting with the mayor and the director of the local tourism organization.

After the transcription of the interviews, the data were analyzed with GABEK (Holistic Processing of Complexity), an analytical strategy to reduce the complexity of texts and to organize knowledge. The analysis was applied with the software WinRelan®, a tool developed to facilitate and support the textual analysis of GABEK (Buber & Zelger, 2000). After splitting all interview transcripts into multiple paragraphs, reflecting single semantic units, the data were manually coded with keywords identified in each text unit. After the data-driven coding process, the most frequent codes appearing in the data were transformed into network graphs (association graphs). These help to organize and understand the landscape of knowledge and opinions of the qualitative data gathered, translating it into patterns and explanations. The graphs help to visualize the interview content and the semantic links of the interview statements.

## *A story about South Tyrol's tourism-migrant nexus*

South Tyrol is a small Alpine region in the Northern part of Italy with around 530,000 inhabitants (Figure 1). Because of the annexation of South Tyrol to Italy after the First World War the region has a long history of cohabitation of cultures, in particular with the Austrian, Italian, and Ladin cultures. The province is considered a good example of the successful protection of linguistic minorities, where cohabitation is based on a complex and differentiated legal system, parity committee formation and proportional representation of all language groups. According to the census data of 2011, 69.4% of the inhabitants speak German, 26.1% Italian and 4.5% speak Ladin, a Rhaeto-Romance language (Astat, 2018a). However, since the nineties new population groups have also become a vital part of society in South Tyrol. In 2018, 9.5 percent of residents had no Italian citizenship (Provincial Statistics Institute [Astat], 2019). This is a higher percentage than, for example, the smallest language group, the Ladins. Persons with a migrant background now represent 'new' minorities in South Tyrol. Until the 1960s, South Tyrol was a region of emigration. Due to the lack of employment opportunities lots of South Tyrolean's migrated foremost to Germany, Switzerland, and Austria. The first strong wave of immigration was when war refugees from ex-Yugoslavia came to South Tyrol. From the 1990s until 2002 the majority of migrants into the province came from the Balkan states and Albania, exceeding those arriving from Western

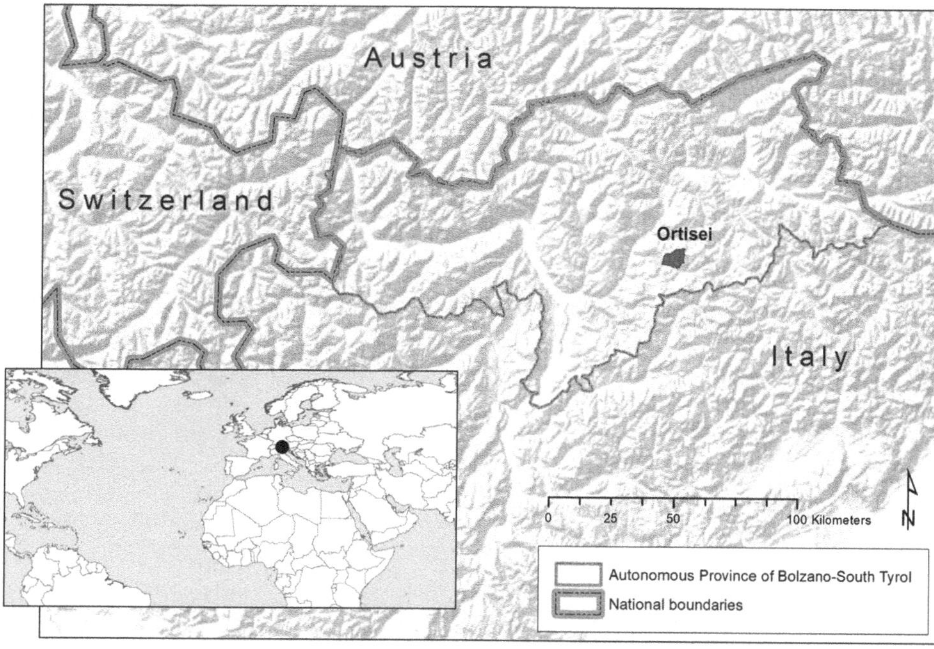


**Figure 1.** Geographical location of South Tyrol and Ortisei (Source: Authors).

Europe. The number of West-Europeans coming to South Tyrol only increased again after 2005.

In contrast to the short immigration history of the region, its history as a tourism destination dates back to the 19th century, when the region was still part of the Austro-Hungarian Empire and aristocrats discovered the Alps for traveling. But the first signs of an increase in tourism numbers became evident after the Second World War, in the 1950s, when a steadily growing wider circle of the population was able to travel and winter sports boomed (Reso, 2016, p.44). South Tyrol was generally characterized by economic growth with the tourism sector becoming increasingly important since the 1970s (Romeo, 2012). In recent decades, competition with other destinations grew, especially due to an increased interest in long-distance travel opportunities, which made further investment and a new market positioning necessary. This included the construction of wellness facilities, expansion of ski operations, and a new image as a destination representing the symbiosis of the Italian and German cultures (Rohrer, 2003). Today, the tourism industry contributes 13.8% to the regional GDP (Astat, 2018a) and is of growing importance for the regional economy (Kofler et al., 2018). In 2018, the region registered about 7.5 million arrivals and more than 33 million overnight stays, mostly domestic visitors or visitors from Germany (Astat, 2019). The accommodation sector consists of around 10,000 businesses with about 225,000 beds (Astat, 2019), mostly small and/or family businesses. According to the statistical office of the European Union (Eurostat, 2019), in 2017, as in the years before, the region was also one of the top 20 tourist destinations in Europe.

The growing importance of the tourism industry, the continuing need for manpower in the sector (Unioncamere, 2019) and its high share of foreign labor illustrate

**Table 1.** Relations between South Tyrol's foreign residents and tourism indicators (Source: Authors).

| | | Number of migrants | | | |
|---|---|---|---|---|---|
| | | *Spearman correlation* | | *Pearson correlation* | |
| Tourism indicators | | $n_1 = 116$ | $n_2 = 114$ (excluding outliers) | $n_1 = 116$ | $n_2 = 114$ (excluding outliers) |
| Number of tourism enterprises | Correlation Coefficient | .618** | .637** | .414** | .757** |
| | Sig. (2-tailed) | 0.000 | 0.000 | 0.000 | 0.000 |
| | N | 114 | 116 | 114 | 116 |
| Number of accommodation facilities | Correlation Coefficient | .465** | .483** | .265** | .165* |
| | Sig. (2-tailed) | 0.000 | 0.000 | 0.002 | 0.038 |
| | N | 114 | 116 | 114 | 116 |
| Accommodation density (accommodations/ km$^2$) | Correlation Coefficient | .371** | .395** | .188* | .230** |
| | Sig. (2-tailed) | 0.000 | 0.000 | 0.022 | 0.006 |
| | N | 114 | 116 | 114 | 116 |
| Number of beds | Correlation Coefficient | .540** | .557** | .273** | .217** |
| | Sig. (2-tailed) | 0.000 | 0.000 | 0.002 | 0.010 |
| | N | 114 | 116 | 114 | 116 |
| Bed occupation rate (%) | Correlation Coefficient | .306** | .334** | .164* | .181* |
| | Sig. (2-tailed) | 0.000 | 0.000 | 0.041 | 0.026 |
| | N | 114 | 116 | 114 | 116 |
| Arrivals | Correlation Coefficient | .561** | .583** | .386** | .542** |
| | Sig. (2-tailed) | 0.000 | 0.000 | 0.000 | 0.000 |
| | N | 114 | 116 | 114 | 116 |
| Domestic arrivals | Correlation Coefficient | .452** | .479** | .232** | .440** |
| | Sig. (2-tailed) | 0.000 | 0.000 | 0.006 | 0.000 |
| | N | 114 | 116 | 114 | 116 |
| International arrivals | Correlation Coefficient | .577** | .599** | .435** | .541** |
| | Sig. (2-tailed) | 0.000 | 0.000 | 0.000 | 0.000 |
| | N | 114 | 116 | 114 | 116 |
| Number of overnight stays | Correlation Coefficient | .517** | .537** | .284** | .268** |
| | Sig. (2-tailed) | 0.000 | 0.000 | 0.001 | 0.002 |
| | N | 114 | 116 | 114 | 116 |
| Number of employed persons in the local tourism sector (%) | Correlation Coefficient | −0.119 | −0.133 | −0.112 | −0.087 |
| | Sig. (2-tailed) | 0.103 | 0.077 | 0.117 | 0.176 |
| | N | 114 | 116 | 114 | 116 |

**Correlation is significant at the 0.01 level (two-tailed).
*Correlation is significant at the 0.05 level (two-tailed).

that tourism and migration are interconnected. In 2018, about 13.9% of employees were working in the regional tourism industry, with a high share of foreign labor (34.7%) (Department for Labor Market Observation, 2018). In South Tyrol, the tourism sector absorbs a large section of the foreign labor force, second only to agriculture. The results of our secondary data analysis show a positive correlation between the number of migrants and almost all tourism indicators: The number of tourism enterprises, beds and overnight stays, as well as the arrivals registered in each municipality, have a high correlation with the number of migrants. Table 1 shows the Spearman and Pearson correlation coefficients of the number of migrants in South Tyrolean municipalities and several available tourism indicators.

The first result to be noted is that both migration and tourism show strong and significant Spearman correlations, excluding the number of employed persons in the local tourism sector. The arrivals differ between domestic (Italian) tourist arrivals and international arrivals. The number of migrants has a higher correlation with international arrivals. The Pearson correlation values indicate a higher nexus between the number of enterprises and migrants ($r = 0.757$), when the outliers Bolzano and Merano were excluded. The correlation coefficients indicate that there is a positive correlation between the number of non-Italian citizens in South Tyrolean municipalities, tourism flows and the local infrastructure. So, a high number of tourism enterprises in a municipality may be an indicator of a higher number of migrants.

In addition to the quantitative study on the nexus of migration and tourism, we wanted to have a deeper look at what a high presence of both migrants and tourists means at a local level. Therefore, the next section gives insights into a tourism-intensive municipality in South Tyrol: Ortisei. The explorative research approach offered the possibility to better understand the interaction and connections of migration and tourism on a local scale and from a local perspective.

### *Insights into everyday life in Ortisei – a touristic hot spot in the Dolomites*

Ortisei is located in the Gardena Valley, one of the two Ladin-speaking valleys of the region. Ortisei is, since the end of the 19$^{th}$ century, an attractive tourism destination. It became a famous winter destination due to the construction of its first lifts and a ski school in the 1950s (Municipality of Ortisei, 2017). In 2019, Ortisei counted about 134,000 arrivals and 696,000 overnight stays, mostly domestic tourists or international tourists from German-speaking countries (Astat, 2019). The importance of the tourism industry is also reflected in the number of residents working in tourism: In comparison to the share at the regional level (13.9% in 2018), Ortisei is a municipality with a very high share of inhabitants working in the sector (36.2%).

Along with the steady increase in tourism, the number of foreign citizens living in Ortisei has also increased (Figure 2). In 2018, Ortisei counted 4,870 inhabitants, of which 9.06% were not Italian citizens. Most have a European passport: 35.0% were from the European Union (EU-28) and 29.5% from other European countries. Figure 2 shows the increase in the absolute number in recent years.

Even though there is no statistical data available on the share of foreign tourism labor at the municipal level, our key informants (experts in the field of tourism and local municipal administration) confirm a large increase in recent years, probably similar to that at the regional level of South Tyrol, where the share increased from 5% in 1998 to 34.7% twenty years later. A stakeholder from the local tourism association (destination management organization) explained: 'Nowadays we have more migrant workers than we had 15 years ago. We need them. At that time, it was viewed as much more critical than today. I believe that the small village of Ortisei has become globalized. [...] We, the tourism industry, called them because we need them.' This statement corresponds to the persistently high demand for and shortages of labor in the industry in recent years (Unioncamere, 2019). On the one hand, the industry has been recruiting seasonal workers from abroad since the 1990s (Rohrer, 2003) and on the other hand, people with a migrant background already living in the area find it easier to get work in tourism than in other sectors (Ladkin, 2018).

Our key informants were asked to describe the impact and role of tourism for Ortisei. Figure 3 illustrates semantic links with Ortisei as a place, the location is

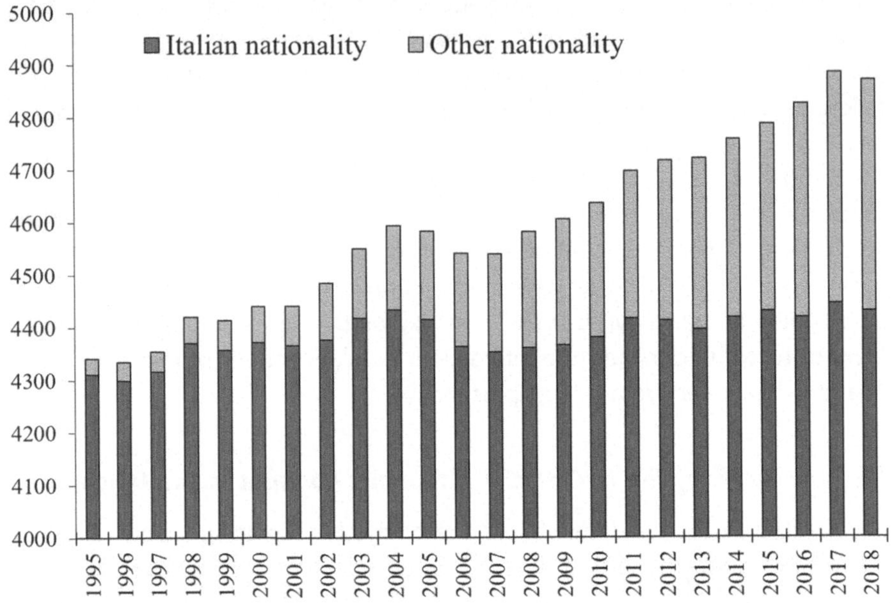


**Figure 2.** Population change in Ortisei (Source: Authors, data: Astat, 2019).

connected to keywords such as family, tourism, and openness. The social advisor of the village, for example, described the impact of tourism on the local population as follows: 'Now with all the people coming through tourism, the population became more open. People with different languages are coming from different parts of the world. Therefore, people from Ortisei [...] had to open up.' Or 'A very important factor in Ortisei is the social cohesion among the population and the tourism stakeholders. They are open-minded and are in contact with others.' These statements indicate that tourism could effectively have an impact on the population's attitude and tolerance toward people from outside their local community.

The findings also show that the interviewees distinguish between locals, tourists, migrants (referring to people moved to the village from abroad) and seasonal workers. It shows that despite the aforementioned openness regarding 'others', the local population makes a clear distinction between themselves and other categories, such as second-home owners, migrants, seasonal workers, tourists, etc. and that the local tourism industry represents an important link between those groups. However, sometimes no distinction was made between foreign residents working in tourism and seasonal workers from abroad. Therefore, the statements show not only the connection between tourism and migration at an economic level, between migrant tourism labor and tourists (host-guest-relationships) or between the tourism industry and migrant workers, they also point to a connection of both phenomena in relation to the residents' acceptance linked to the length of or type of stay by the aforementioned actors, as emphasized by the social advisor: 'They came here mostly because of their work and then stayed there, started a family and then lived with us. That's already 10% of the population in Ortisei.' The integration advisor commented: 'Tourists are welcome. And they [the locals] say about the migrant workers: "As long as they have

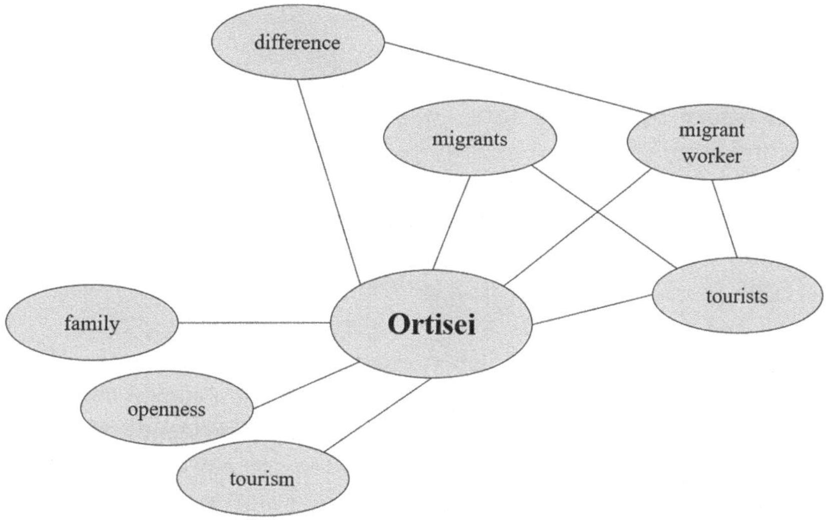


**Figure 3.** Semantic links with the keyword 'Ortisei' (Source: Authors).

a job, it is okay."' But the findings also show that some groups tend to create relationships within the same background, as the Peruvian group illustrates, according to the social advisor: 'Some of them [migrant-workers] will certainly build relationships, but the reality is that the Peruvians, for example, are staying among themselves. They've even founded an association. They make their celebrations and have their own network.' Encounters between the different population groups living in Ortisei happen, according to the social advisor, either via the workplace or the second or third generation: '[...] because they have been here for a long time and now also have children that go to school.' The example of the Peruvian community shows that some communities have their own way of life, that each group is differently integrated into the life of the community of Ortisei, and that over time, they're also seen as part of the local society.

A closer look at the terms 'migrants' and 'tourists' reveals that people make a clear distinction between these groups (Figure 4), even if sometimes the differences blur. The duration or length of stay is one way to distinguish between both groups, but the reason for their stay is the most important one: Tourists are visiting the destination for leisure and holidays, they are consumers and customers and stay for a limited period. In contrast, migrants are immigrants, new population groups, or foreigners, who come for work and to earn money, intending to stay for a season or longer. Migrants are mostly seen as an additional workforce that helps to compensate for labor shortages. The interviewed tourism entrepreneur commented: '[...] there are differences between migrants and tourists. A tourist is someone who comes here to spend his holiday.' And the migrant worker stated: 'The difference is that a foreigner comes here for work, a tourist is on holiday.' These statements highlight the importance of an impelling motive to move as a central defining feature.

In this study, the term 'tourist' was linked with several European countries, such as Germany, Italy, England, the Czech Republic, and Russia; countries that represent important source markets for travel to the destination Ortisei. Compared to the rest of

South Tyrol, it has a more international tourist audience. Other terms related to tourists highlight their role as customers, consuming tourism products or experiences related to the environment such as 'mountains' or the 'Dolomites' (see Figure 4). The cultural and natural settings are also central aspects that help to attract established and new visitor segments.

In contrast, migrants were linked with concepts such as family, kindergarten, school, locals, and work (see Figures 3 and 4). These notions indicate a long-term stay, even though a lot of migrant workers are seasonal workers and stay just for a limited time in Ortisei. Due to their long-term stay (compared to tourists), they were associated with systems and structures relevant for residents and locals. Despite the clear separation between native and new population groups highlighted earlier, questions of the long-term inclusion of new citizens become more visible. People coming from Peru were often mentioned in this context since many stay permanently. Nonetheless, the interviewees made also a clear distinction between voluntary and forced migration and referred to the asylum seekers and refugees living in the valley, due to the increased refugee flows of the past years, especially in 2015. The high demand for tourism labor sometimes made it easier for refugees, too, to find work in the local context. Generally, for refugees the job search is very different, as they still have to learn the language or because they have a different educational background (Marcher et al., 2017).

The interviewees divided migrants into people that come from 'far away' and from geographically 'closer' nations. Migrant tourism workers come predominantly from Central and Eastern European countries, such as Slovakia, and come to the region for seasonal work and are therefore considered as short-term migrants from 'closer' regions. Refugees and people from Peru, and also those from Pakistan, are considered as long-term migrants. They stay in Ortisei and don't go back to their country of origin ('stay') between the seasons. They settled and live with their families in the village all year round.

Moreover, our informants also distinguished two different groups of tourists: regular guests and second-home owners. The traditional guests or regular visitors are those who come every year and get to know the residents. They are familiar with the destination and appreciate the professional service quality. Over the years they often developed a personal relationship with hoteliers and/or other familiar residents. One interviewee described them as follows: 'They [regular guests] would like to be like locals, but they simply aren't. They feel like stepchildren of the situation here. Not all of them, but these guests are different than the normal ones. Maybe because they are more demanding.' The second group of tourists is the group of second-home owners mostly from other Italian regions, who visit the village several times a year. According to the interviewees, they also have a close relationship with the village and its people. But, the attitude of locals toward them is rather negative. They perceive second-home owners as those responsible for the increase of property and rental prices. One of the interviewees described second-home owners as 'neither guests nor locals.' This shows that even though there seem to be two clearly distinguishable groups of mobile people (migrants and tourists), some type of 'movers' challenge not only the residents' definition of who is a migrant or a tourist, but also who is a 'local.'

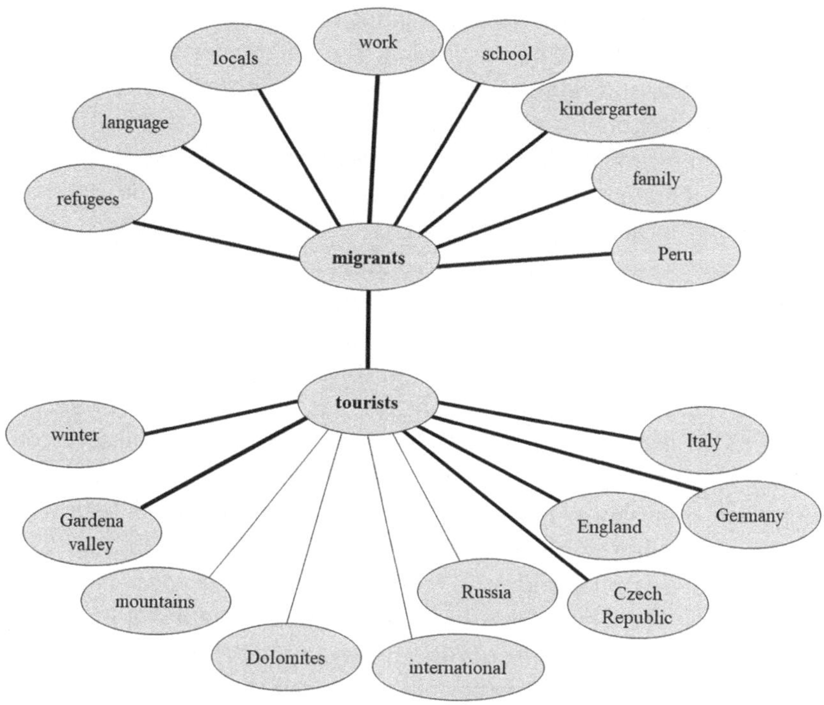


**Figure 4.** Migrants versus tourists: categorization through semantic connections (Source: Authors).

Despite the differences between the two groups, one central term that appeared very often was 'trust,' highlighting the relational aspects between locals, migrants, and tourists. According to the interviewees, all relations and interactions between the different actors within the municipality – migrants, locals, and tourists – are characterized by different levels of trust. Trust between those different groups can be created and supported through social interaction, gained in different ways, by work, knowledge or experience. Trust can be built through social interactions or relations, but also represents a necessary factor for strong, mutual, social relationships. Much literature deals with the concept of trust (Humphrey & Schmitz, 1996; Newton, 2001; Nooteboom, 2007) and prior research in South Tyrol has already shown the strong impact of trust on the inclusion of new population groups, such as asylum seekers and refugees (see Kofler & Marcher, 2017; Marcher et al., 2017). But several studies have shown that economic inequality and linguistic diversity reduce trust (Boix & Posner, 1998; Brockner & Siegel, 1996; Park & Subramanian, 2012). According to these arguments, the level of trust in Ortisei should be considered as rather low as the economic situations of locals, tourists, and Peruvian migrant workers are very different.

## Conclusion

Still, at first glance, national and statistical categories seem to be central from a local perspective to explain how migration and tourism have affected the village and which new groups of actors have become a part of the local reality or influence it. So, tourists and migrants, especially refugees, cannot be considered as equal, although all of them may end up in the same destination. The differences become more evident regarding legal aspects and rights as well as the motivation to stay or visit the destination, even if the categories sometimes blur regarding the length of stay and the underlying spatial movement. Regulations and definitions established on different spatial scales build, therefore, a general framework and impact the conditions and cohabitation at the local level. Regular guests or second homeowners, for example, are not seen as part of the community in contrast to migrants living in the area for a long time. Another aspect in which migrants and tourists differ in Ortisei is their country of origin, because different countries provide different possibilities. This is often referred to as the lottery of the place of birth (Hillman, 2016). The case of Ortisei has shown that the most current tourist groups are not from the same country of origin as the migrants working in the tourism industry. While tourists mainly come from Western Europe, Germany, Great Britain or Italy, and more and more also come from Eastern European countries or Russia, for the interviewees, migrants seem to be most likely from Peru. People from Central or Eastern European countries, such as Slovakia, are also important (seasonal and long-term) workers in the local tourism industry. According to our findings, one nexus between the two phenomena is, therefore, that the groups belonging to both are not local, they are different to the native inhabitants; other commonalities are that they are connected through the industry and, last but not least, all of them somehow interact with the local population and, thus, depend on the support of the 'hosting' society.

The case study presented shows, in part, what researchers define as tourism-led migration (Williams & Hall, 2000), which refers to the fact that tourism contributes to the creation of labor migration. Especially in a region with a high shortage of (skilled) workers in the tourism industry, such as South Tyrol (WIFO, 2018), migrants are an important group among the labor force and are also attracted by the tourism labor market. The qualitative interviews with local key informants underlined this argument, even though it strongly depends on their level of agency and their professional or migrational background. As other research showed, the labor market access for migrants, especially for refugees, is often extremely difficult (Marcher et al., 2017). For example, the search for work depends on many different factors, ranging from the educational background of the individual to the perception of new population groups by the host society. Additionally, the case of Ortisei has shown that migration leads to tourism. The fact that migrant workers stay away from their homes for long periods of time motivates their friends and relatives to visit them. Thus, for example, Peruvian migrants in Ortisei trigger trips that otherwise would not have occurred. Based on these mechanisms, the correlation between the number of migrants living in the region and the selected tourism indicators was foreseeable (see also Massidda et al., 2017).

Tourists and migrants make important social and economic contributions to destination countries, and they culturally enrich society. The presence of migrants can also provide a destination with more diverse and tourist-related offerings for visitors, which is more evident in cities (e.g. with districts such as 'Little Italy' or 'Chinatown') (Dwyer et al., 2014, p. 131). Migrants may not only provide the resource 'labor' for the travel, tourism, hospitality, and catering sectors and generate visits through friends and relatives, they may also influence local individual and collective values, norms and lifestyles and may enhance the tourism product. On the one hand, the case of Ortisei has shown that tourism and, thus, the presence of people from different countries and with different nationalities, has contributed to more 'openness'. And on the other hand, the findings illustrated that there could be a tendency of residents of tourism-intensive destinations with a critical mass of migrants to be more open-minded and tolerant, probably more than residents of destinations with less tourism and migration. However, further research and investigations are needed to underline this thesis.

People with migration backgrounds who have been in Ortisei for a very long time seem to have become a part of the community and everyday life. Some parallels appear also with regular guests and tourists with a long duration of stay. They increase the level of trust between the 'we' and the 'stranger' and seem to give a sense of security through regular and repeated stays. Certain regularities and the duration of stay seem to change and affect the perception of the 'other' and the level of trust between certain groups, although the attitude of locals toward short-term and regular visitors is more positive than their attitude toward second-home owners. The latter are held responsible for higher rents and land prices. At the most basic level, the relationship between tourism and migration can be viewed as a spectrum with permanent migration at one extreme, and tourism at the other (Bell & Ward, 2000). An additional explanation for maintaining a certain distance from 'others', whether they be a tourist or a migrant, may be that 'locals' of minority areas, such as South Tyrol, possess a very strong collective identity and a higher awareness of their cultural and linguistic particularities.

Authors, such as López Martín (2017), analyzed the effects of human mobility on generalized trust both through short-term and long-term relations. She emphasizes that migration has a positive impact on generalized trust. In the context of Ortisei, it would also be interesting to consider studies from management or network theories as well as social capital (Bourdieu, 2012; Putnam, 2001; Putnam et al., 1994) focusing on the relationship between trust and mutual dependencies and to observe if the level of trust changes over time with the level of dependency. However, this would be a research question for further investigations.

This case study has shown that interactions, permanent exchange, and the complex interrelationships of actors that seem widely different at first sight may, in the long run, create trust and familiarity between the different population groups. But for a deeper understanding of the tourism-migration nexus, more in-depth investigations of the municipality concerned and a larger and more diverse sample would be necessary. The same research design could be applied to destinations with a long history of tourism and destinations with less tourism in different parts of the world. Another interesting aspect for further studies would be a historical perspective, to better understand developments over time.

The connections between migration and tourism at the local and regional level are manifold, ranging from economic incentives to questions about cohabitation and the structure of everyday life. Nevertheless, their examination can contribute to a better understanding of a small fraction of the complex, global interdependencies of our time.

## Disclosure statement

No potential conflict of interest was reported by the authors.

## ORCID

Ingrid Kofler http://orcid.org/0000-0002-9743-3122

## References

Astat [Provincial Statistics Institute]. (2018a). *Database*. Retrieved January 20, 2018, from http://astat.provinz.bz.it/de/datenbanken-gemeindedatenblatt.asp

Astat [Provincial Statistics Institute]. (2018b). *Zeitreihe des Tourismus/Serie storica sul turismo 1950-2017* [Tourism time series 1950-2017]. (Report No. 9.). http://astat.provinz.bz.it/de/mobilitaet-tourismus.asp

Astat [Provincial Statistics Institute]. (2019). *Database [online]*. Retrieved February 14, 2019, from https://astat.provinz.bz.it/de/datenbanken-gemeindedatenblatt.asp

Bell, M., & Ward, G. (2000). Comparing permanent migration with temporary mobility. *Tourism Geographies, 2* (1), 87–107. https://doi.org/10.1080/146166800363466

Boix, C., & Posner, D. N. (1998). Social capital: Explaining its origins and effects on government performance. *British Journal of Political Science, 28*(4), 686–693. https://doi.org/10.1017/S0007123498000313

Bourdieu, P. (2012). Ökonomisches Kapital, kulturelles Kapital, soziales Kapital [Economic capital, cultural capital, social capital]. In U. Bauer, U. H. Bittlingmayer, & A. Scherr (Ed.), *Handbuch Bildungs- und Erziehungssoziologie* (pp. 229–242). VS Verlag für Sozialwissenschaften.

Brockner, J., & Siegel, P. (1996). Understanding the interaction between procedural and distributive justice. In R. M. Kramer & T. R. Tyler (Ed.), *Trust in organizations* (pp. 390–413). Sage.

Buber, R., & Zelger, J. (2000). *GABEK II: Zur qualitativen Forschung [GABEK II: For qualitative research]*. Studien Verlag.

Department for Labor Market Observation. (2018). *Labor market data online*. Bolzano-Bozen, Autonomous Province of Bolzano-Bozen. http://www.provinz.bz.it/arbeit-wirtschaft/arbeit/statistik/arbeitsmarktdaten-online.asp

Dwyer, L., Seetaram, N., Forsyth, P., & King, B. (2014). Is the migration-tourism relationship only about VFR? *Annals of Tourism Research*, *46*, 130–143. https://doi.org/10.1016/j.annals.2014.03.009

Etzo, I., Massidda, C., & Piras, R. (2014). Migration and outbound tourism: Evidence from Italy. *Annals of Tourism Research*, *48*, 235–249. https://doi.org/10.1016/j.annals.2014.07.002

Eurostat. (2019). *Eurostat regional yearbook 2019*. European Union. https://ec.europa.eu/eurostat/en/web/products-statistical-books/-/KS-HA-19-001

Friese, H. (2014). *Grenzen der Gastfreundschaft: Die Bootsflüchtlinge von Lampedusa und die europäische Frage. [Limits of hospitality: The boat refugees of Lampedusa and the European question]*. Transcript Verlag.

Glasze, G., & Pott, A. (2014). Räume der Migration und der Migrationsforschung [Spaces of migration and migration studies]. In P. Gans (Ed.) *Räumliche Auswirkungen der internationalen Migration: Forschungsberichte der ARL* (Vol. 3, pp.47–62). DE: ARL-Verlag. https://www.ssoar.info/ssoar/handle/document/43884

Glick Schiller, N., & Salazar, N. B. (2013). Regimes of mobility across the globe. *Journal of Ethnic and Migration Studies*, *39*(2), 183–200. https://doi.org/10.1080/1369183X.2013.723253

Hall, C. M. (2005). *Tourism. Rethinking the social science of mobility*. Pearson Education Limited.

Hillman, F. (2016). *Migration. Eine Einführung aus sozialgeographischer Perspektive*. Franz Steiner Verlag.

Holert, T., & Terkessidis, M. (2006). *Fliehkraft: Gesellschaft in Bewegung - von Migranten und Touristen. [Centrifugal force: Society on the move – From migrants and tourists]*. DE: KiWi-Taschenbuch.

Humphrey, J., & Schmitz, H. (1996). *Trust and economic development* (IDS Discussion Paper 355). Institute of Development Studies.

Illés, S., & Michalkó, G. (2008). Relationships between international tourism and migration in Hungary: Tourism flows and foreign property ownership. *Tourism Geographies*, *10*(1), 98–118. https://doi.org/10.1080/14616680701825271

Kalir, B. (2013). Moving subjects, stagnant paradigms: Can the 'mobilities paradigm' transcend methodological nationalism? *Journal of Ethnic and Migration Studies*, *39*(2), 311–327. https://doi.org/10.1080/1369183X.2013.723260

Kofler, I., & Marcher, A. (2017). Immigrazione e reti sociali nelle aree montane: Due casi studio di integrazione di richiedenti asilo [Immigration and social networks in mountain areas: Two case studies of asylum seekers' integration]. In A. Membretti, I. Kofler, & P. P. Viazzo (Eds.), *Per forza o per scelta* (pp. 175–185). IT: Aracme Editrice.

Kofler, I., Marcher, A., Pechlaner, H., & Volgger, M. (2018). The special characteristics of tourism innovation networks: The case of the Regional Innovation System in South Tyrol. *Journal of Hospitality and Tourism Management*, *37*, 68–75. https://doi.org/10.1016/j.jhtm.2018.09.004

Kopnina, H. (2007). *Migration and tourism: Formation of new social classes. Tourism dynamics*. Cognizant Communication Offices.

Koslowski, R. (2018). Think mobility instead of migration: Leveraging visitors, tourists and students for more international cooperation. *Columbia Journal of Transnational Law*, *56*(2), 263–273.

Ladkin, A. (2018). Tourism human resources. In C. P. Cooper, S. Volo, & W. C. Gartner (Eds.), *The SAGE handbook of tourism management* (pp. 254–269). SAGE Publications.

Lenz, R. (2010). *Mobilitäten in Europa: Migration und Tourismus auf Kreta und Zypern im Kontext des europäischen Grenzregimes* [Moblitities in Europe: Migration and tourism in Crete and Cyprus in the context of the European Border Regime] [Doctoral dissertation]. VS Verlag für Sozialwissenschaften.

López Martín, C. S. (2017). *International migration and tourism: Determinants of generalized trust in developing countries* [Master's thesis]. https://pdfs.semanticscholar.org/d9d6/a0b2028836648ada0504d233141abb442003.pdf

Marcher, A., Kofler, I., & Streifeneder, P. T. (2017). Social and labor integration of asylum seekers in rural mountain areas—A qualitative study [Special issue]. *Mountain Research and Development*, *37*(4), 388–395. https://doi.org/10.1659/MRD-JOURNAL-D-17-00030.1

Massidda, C., Etzo, I., & Piras, R. (2017). The relationship between immigration and tourism firms. *Tourism Economics, 23* (8), 1537–1552. https://doi.org/10.1177/1354816617708660

Municipality of Ortisei. (2017). *St. Ulrich stellt sich vor [St. Ulrich introduces itself]*. Retrieved October 11, 2017, from http://www.comune.ortisei.bz.it/system/web/zusatzseite.aspx?menuonr=219389085&detailonr=217478552

Newton, K. (2001). Trust, social capital, civil society and democracy. *International Political Science Review, 22*(2), 201–214. https://doi.org/10.1177/0192512101222004

Nooteboom, B. (2007). Social capital, institutions and trust. *Review of Social Economy, 65*(1), 29–53. http://www.jstor.org/stable/29770394 https://doi.org/10.1080/00346760601132154

Park, C., & Subramanian, S. V. (2012). Voluntary association membership and social cleavages: A micro-macro link in generalized trust. *Social Forces, 90*(4), 1183–1205. https://doi.org/10.1093/sf/sos059

Putnam, R. D. (2001). *Bowling alone: The collapse and revival of American community.* Touchstone, Simon & Schuster.

Putnam, R. D., Leonardi, R., & Nanetti, R. Y. (1994). *Making democracy work: Civic traditions in modern Italy.* Princeton University Press.

Reso, E. (2016). *Das große Ganze: Intergenerationalität in familiengeführten Tourismusbetrieben in Südtirol [The big picture: Intergenerationality in family-owned tourism firms in South Tyrol]* [Doctoral dissertation]. Waxmann.

Rode, N. (2017). *The tourism-migration nexus: Towards a theory of global human mobility* [Master's thesis]. https://digital.library.ryerson.ca/islandora/object/RULA:582

Rohrer, J. (2003). Das Produkt Südtirol [The product South Tyrol]. In G. Solderer (Ed.), *Das 20. Jahrhundert in Südtirol. Zwischen Europa und Provinz* (pp. 204–217). IT: Edition Raetia.

Romeo, C. (2012). *Tirol Südtirol Trentino: Ein historischer Überblick/Tirolo Alto Adige Trentino: Uno sguardo storico. [Tyrol South Tyrol Trentino: A historical overview]*. GECT Euregio Tirolo Alto Adige Trentino.

Salazar, N. B. (2018). *Momentous mobilities*. Berghahn Books.

Seetaram, N., & Dwyer, L. (2009). Immigration and tourism demand in Australia: A panel data analysis. *Anatolia, 20*(1), 212–222. https://doi.org/10.1080/13032917.2009.10518905

Sheller, M., & Urry, J. (2016a). The new mobilities paradigm. *Environment and Planning A: Economy and Space, 38*(2), 207–226. https://doi.org/10.1068/a37268

Sheller, M., & Urry, J. (2016b). Mobilizing the new mobilities paradigm. *Applied Mobilities, 1*(1), 10–25. https://doi.org/10.1080/23800127.2016.1151216

Unioncamere. (2019). *Bollettini annuali*. https://excelsior.unioncamere.net

Urry, J. (2007). *Mobilities*. Polity Press.

WIFO. (2018, January). *Fachkräftemangel: Herausforderung für Südtirol [Skills shortage: A challenge for South Tyrol]*. Retrieved February 18, 2019, from http://www.bz.camcom.it/de/fachkr%C3%A4ftemangel-herausforderung-f%C3%BCr-s%C3%BCdtirol

Williams, A. M., & Hall, C. M. (2000). Tourism and migration: New relationships between production and consumption. *Tourism Geographies, 2*(1), 5–27. https://doi.org/10.1080/146166800363420

# Local gastronomy, transnational labour: Farm-to-table tourism and migrant agricultural workers in Niagara-on-the-Lake, Canada

Kristin Lozanski and Kayla Baumgartner

**ABSTRACT**

Each year, migrant workers from Mexico and the Caribbean travel to Canada via a bilateral agreements to provide labour essential to the agricultural sector. One destination for these workers is Niagara-on-the-Lake (NOTL), a microclimate in which labour intensive crops such as tender fruits and grapes are grown. These local crops form the basis of a significant gastrotourism industry. This industry turns on the narrative of an historical idyll in which producer and consumer share a close relationship, one manifested in the farm-to-table discourse that permeates NOTL. Yet agricultural production in the area is entirely dependent upon a globalized labour force, a dependence that is inconsistent with the narrative of locality. The structure of this transnational labour program and the requisite aesthetics of tourism in NOTL render migrant workers and their labour invisibility and this invisibility, in turn, exacerbates the precarity of transnational workers within global capitalism.

摘要

每年，来自墨西哥和加勒比的外来工人通过一项双边协定前往加拿大，为农业部门提供必要的劳动力。这些工人的一个目的地是尼亚加拉湖上地区(NOTL)，这里是一种小气候，适宜种植鲜果和葡萄等劳动密集型作物。这些当地作物构成了重要的美食旅游业的基础。这一产业开启了一种历史田园诗般的叙事，生产者和消费者之间有着密切的关系，这种关系在弥漫于尼亚加拉湖上地区的"从农场到餐桌"的话语中得到了体现。然而，该地区的农业生产完全依赖于全球化的劳动力，这种依赖与地方性的叙述不尽一致。这种跨国劳工计划的结构和旅游业所必需的审美观使移徙工人及其劳动难以呈现，而这种跨国劳动的隐形又加剧了跨国工人在全球资本主义中的不稳定性。

As an August sunset falls over the quiet town of Niagara-on-the-Lake, Ontario, Canada, groups of guests begin to gather around large picnic tables set up on a street temporarily closed to car traffic for the final event of the Shades of Summer Festival. Upon each group's arrival, they begin adorning the tables with props and decorations ranging from balloons and flowers to rainbow leis and a palm tree made of pineapples. Attendees are bedecked in crisp, white dress clothes, consistent with the event's dress code. They carry with them an array of food and coolers to stash their wine and beer. Without a speech to formally mark the start of the ceremony, they begin indulging as soon as they finish setting up. A DJ puts on a rotation of soft-rock hits, and the 200-odd attendees spend the evening eating, drinking, chatting, and dancing.

## Peaches on queen street: Agritourism in niagara-on-the-Lake

The Shades of Summer dinner caps off Niagara's Peach Festival, an annual weekend-long celebration of the local fruit harvest in mid-August. The event showcases many of the region's agricultural specialties. Throughout the weekend, vendors appear on the streets selling peach pie, peach jam, and even peach ice cream, along with peaches themselves. Visitors can also find locally harvested nectarines, plums, strawberries, cherries, and apricots. Locally produced wine receives significant attention, though sales are restricted to the evening. The Shades of Summer capstone event is open to everyone, provided they pay $35 per person, or $240 for a table of eight. Dinner and drinks are not included—guests must bring their own, but waitstaff are on hand to help with serving. Festival rules state that patrons must purchase wine (VQA[1] only) and beer from the event organizers, but enforcement is lax. As the night winds to a close, festival organizers begin clearing away the tables, chairs, and all other evidence of the night's festivities. With the final removal of the roadblocks, the street once again reverts to a public thoroughfare.

Conspicuously absent from the dinner were those who actually harvested the peaches. Through Canada's Seasonal Agricultural Worker Program (SAWP), migrant farm workers arrive in Niagara-on-the-Lake (NOTL) every year beginning in March to work the fields, earlier if they are working in greenhouses. They come from Mexico and the Caribbean[2], labouring on Canadian farms for up to eight months[3]; the Niagara region – a micro-climate that enables the production of tender fruit and grape crops – hosts approximately 3 000 workers through the SAWP annually. Workers have been coming to Canada through SAWP for more than 50 years. These 'temporary' workers are vital to the Canadian economy. In 2016, over 40 000 workers (Employment & Social Development Canada, 2018a) coming from Mexico and 11 Caribbean countries to support an industry that contributed $111.9 billion – 6.7% – to Canada's GDP (Agriculture & Agri-Food Canada, 2017). These workers are rendered invisible not only in such public celebrations of local agriculture and harvest, but also in their daily labour as their presence challenges the narrative of NOTL's local – and implicitly ethical – production of food. While the local food may be produced within a relatively small radius, the presence of workers from Mexico and the Caribbean signals that this 'local' remains embedded within the practices of global capitalism.

The curated touristic narrative of Niagara-on-the-Lake draws heavily upon local agriculture and harvest not only in its festivals, but as a mainstay of its tourism storyline. Yet the living and working conditions of the workers who sustain agriculture, and the workers themselves, are inconsistent with the tourist gaze (Urry, 1990). Labour is discordant with a tourist gaze because 'tourism is a leisure activity which presupposes its opposite, namely regulated and organized work. It is one manifestation of how work and leisure are organized as separate and regulated spheres of social practice in 'modern' societies' (Urry, 1990, p. 4). Moreover, agricultural labour in NOTL is disruptive of a tourist gaze that desires a pre-modern rural idyll as the racialized bodies of the labourers reflect globalized systems of production.

## *A note on research methods*

This analysis of tourism and agricultural production in Niagara-on-the-Lake (NOTL) draws upon participation observation in winery tours and ethnography in the area generally, but also specifically with farm workers, over four years beginning in the spring of 2016 by both authors[4]. The participant observation of tourism in NOTL has included tours of many of the wineries in the area, during which questions about the role migrant workers were asked; time spent on Queen Street (the main street) of NOTL noting the types of establishments present and observing tourists; eating at the farm-to-table restaurants, with close attention to the framing of the food within the menu and by staff. In addition, pamphlets, booklets, and posters in the tourism information centre have been collected as secondary data.

Kristin Lozanski has ongoing research relationships with several Jamaican men working at a handful of farms in the area and visits them approximately monthly from April through October. These workers speak English and Patois. They facilitate conversations between the researchers and workers who speak primarily Patois. Visits with the workers may include providing transportation for errands or going together to Niagara Falls. More often, though, these visits involve sitting with the workers at their farmhouses and speaking informally with them. Whatsapp is also used for brief communications with workers while they are in Canada – to see how they are doing, to let them know when we will be visiting – and less frequently while they are in Jamaica. Lozanski has also conducted two fieldwork trips to Jamaica (November 2017 and December 2019) visiting workers in their homes, with one of the men she knows best as her guide. Only two formal interviews with workers have been done because of the significant precarity of the workers and the risk associated with participating in this research (see below). During those interviews, the workers were visibly nervous. These men are regarded as leaders by their peers and they are willing to act as a liaison in bringing forward workers' concerns to their employers. Despite their ability to engage critically with farm owners and managers, they told the interviewer that if their employers knew they were participating in an interview, they would be sent home the next day. While other researchers have elected to use interviews as a means to collect data (Basok et al., 2014; Binford, 2013; Hennebry, 2012; Preibisch, 2010), we felt this method would be an impediment to rapport and relationship-building in our specific circumstances. As relationships and data collection are ongoing, we have opted to collect data through informal conversations in which we regularly explain that we are learning about their experiences so that Canadians will better understand the living and working conditions that underpin local food production.

Workers understand that we are collecting data about their experiences. Our positions as university-based researchers has been clearly and repeatedly articulated since beginning the project. Workers' participation is voluntary in so much as they can choose to speak or not speak with the researcher. During our visits, we write notes openly in front of the workers in notebooks; these notes are drafted into extensive field notes as soon as possible after the visit. Workers have read (or have had material read to them) and approved any work that makes reference to their individual experiences or events specific to their farms. These approval processes have included extensive conversations about the risks of being identified, which they understand keenly, and the possibilities for mitigating those risks by changing details of their stories. The workers make the final decisions and approve all changes to, and omissions of, potentially identifying details. In this way, beyond the approval that has been received by our institutional Research Ethics Review Committee, our research ethics are informed by the understanding that

> An ethics protocol is primarily meant to support an ongoing dialogue with vulnerable communities and individuals who are participants in our research, and that considering ethical issues is about more deeply understanding how our participants' vulnerability might position them vis-à-vis the research and us as researchers. (Roger & Mignone, 2018, p. 47)

While the idea of 'research' is quite abstract for many of the workers, they understand that the information they share during our visits is the information that will be included in a book about them and their experiences – the ultimate goal of this research and one with which they can identify. The stories and experiences of the workers are consistent with the excellent and extensive scholarship that has been documenting the experiences of SAWP workers in Ontario for more than a decade (see, as a sample of this scholarship, Basok, 2003, 2004; Binford, 2009, 2013; Hennebry, 2012, 2014; McLaughlin, 2010; Mayell, 2016; Preibisch, 2004, 2010; and Reid-Musson, 2014, 2017).

In this article, we bring together data from our participant observation and ethnographic research to map the dissonance between NOTL's nostalgic discourse of local gastrotourism and the working conditions of those people who tend the crops that constitute the foundation of this touristic narrative.

## *Heritage tourism in NOTL*

Old Town[5] is 22 km from its better-known neighbour – Niagara Falls. While Niagara Falls is a global tourist destination and home not only to the Horseshoe Falls, but also many sites of popular tourism – including its casino, fun houses, chain restaurants, and historical ties to honeymooning and heterosexuality (Dubinsky, 1999) – NOTL has actively produced itself as a more refined destination, 'one intended to be distinct from the mass tourism of Niagara Falls' (Morgan, 2015: 113). The rides, haunted houses, bright lights, and street side speakers in Niagara Falls contrast with the horse-drawn carriages, heritage designations, and Shaw Theatre Festival in NOTL. NOTL claims to be 'Prettiest Town in Canada' (based on winning this title from a Communities in Bloom competition in 1996) (Figure 1).

Through the Niagara Parks Commission, a parastatal organization founded in 1885 to manage the development of public spaces in Niagara Falls (Healy, 2006), NOTL has curated its sophisticated reputation. The mission of the Niagara Parks Commission is to protect the natural and cultural heritage along the Niagara River for the enjoyment

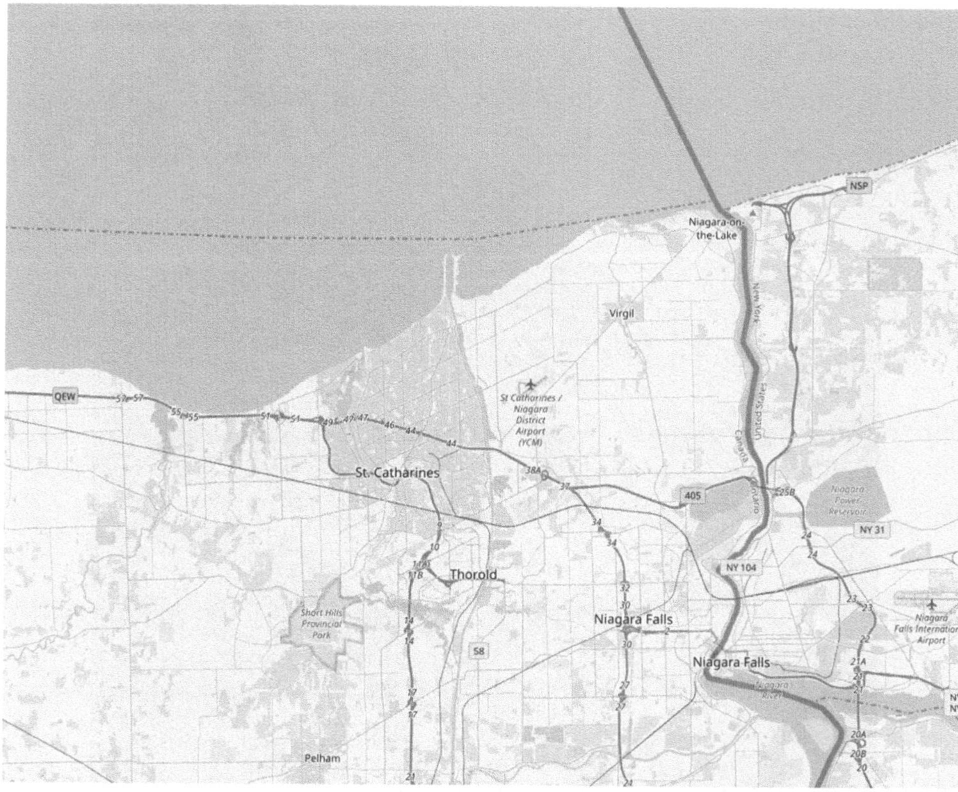


**Figure 1.** Niagara region map. Niagara-on-the-Lake is a microclimate on the south shore of Lake Ontario, bounded on the south by the Niagara Escarpment. The Niagara River, which forms the Canada-United States border in that area, cascades over the Escarpment at Niagara Falls approximately 25 km away. On clear days, the city of Toronto is visible across Lake Ontario.
© OpenStreetMap contributors, License.

of visitors while maintaining financial self-sufficiency. As guardian of a national trust, we are mandated to preserve and commemorate the historical, cultural and environmental significance of the Niagara River corridor (Niagara Parks, 2020).

Unlike Niagara Falls – a destination of 'amusement' – 'the Parks Commission and local historical societies have been active since the 1920s, but especially since the 1960s, refurbishing forts, old buildings, burial grounds and historical markers at NOTL, quite consciously shaping the tourist gaze around education and history' (Dubinsky, 1999, pp. 242–43). It tightly manages both streetscape and crowds through rigorous architectural and parking requirements (Town of Niagara-on-the-Lake, 2013), an obligation for businesses that indirectly, but significantly, restricts the types of retailers able to obtain municipal approval (Crag & Canyon, 2007). The streets are lined with predominantly independent retailers; there are no chain stores or cafes with the exception a small Valu-mart, the grocery store in Old Town (Figure 2).

As a tourist destination, NOTL highlights its heritage as well as its local food and wine production (Telfer & Hashimoto, 2013). Indeed, NOTL submitted an application to Parks Canada, seeking federal nomination as a UNESCO World Heritage Site. The Outstanding Universal Values articulated in the application were the area's cultural

**Figure 2.** Caleche on Queen St. Drawing upon the historical narrative of through which Niagara-on-the-Lake produces itself, a horse-drawn carriage makes its way along Queen St. past the historic Court House Theatre. © Kristin Lozanski

landscape alongside the land use and settlement by vulnerable settlers and cultures. The application highlighted requisite protections within and authenticity of the site: 'The cultural landscape, while still vital and evolving, includes cultural resources with protected heritage values that authentically preserve evidence of the human experience within the landscape,' (Niagara-on-the-Lake Application Submission, 2017).[6] This commitment to heritage sits incongruously alongside the Niagara region (including Niagara Falls and St Catharines) experiencing the fastest rising house prices in Canada from 2016 to 2019 (Bisby, 2019), prices driven in part by the purchase of real estate for short-term rental use (Forsyth, 2018) and in part by recent high end to luxury developments in the area.

Despite the pressures of economic expansion experienced by NOTL residents, NOTL is effectively understood through the lens of heritage tourism in which the past is commodified and brought into the use of the present (Graham et al., 2000; see also Baxter-Moore & Charest, 2013). Nostalgia for the colonial era is evident not only in the overt orientation towards its heritage, but also in current day capitalist ventures. Queen Street – the main street in Old Town – caters to a bourgeois clientele through the provision of tasteful clothing, jewellery, specialty housewares, and food products by independent restaurants and shops. More overtly, some retailers directly recall Canada's British colonial heritage: Irish Design; the Scottish Loft; and The Name Tree, which trades in family histories and coats of arms.

Indeed, NOTL explicitly frames the touristic experience it provides through the language of heritage:

> Our town attracts thousands of visitors who stroll the *old* streets, enjoy the comforts of fine hotels or our *charming* bed and breakfast homes, dine in our many varied restaurants, and, if even for a little while, journey *back in time to a more leisurely time* - the age of Niagara-on-the-Lake! (Tourism Niagara-on-the-Lake, 2016; emphases added)

An emergent discourse within the townsite engages directly with more troubled forms of the area's history. In 2018, the sign at the main entrance to the downtown area changed from 'St. Mark's Church ~ since 1792' to 'Chloe Cooley ~ Act to Limit Slavery ~ 1793.' In November 2018, the Town opened the Voices of Freedom Park to 'commemorate[e] the town's black history and the 222th anniversary of the Act to Limit Slavery in Upper Canada' (Voices of Freedom, 2018). NOTL actively claims its significance in Canadian history as the site in which Lieutenant Governor John Graves Simcoe limited slavery in Upper Canada, ultimately abolishing it, but through a period of two generations (Power & Butler, 1993). Though Simcoe introduced the *Act to Limit Slavery* in 1793, slavery was still practiced in Canada, under the terms of Simcoe's *Act*, until the passage of the *Abolition of Slavery Act* by Britain, effective in 1834 (Henry, 2010).

While NOTL implicitly recognizes slavery as having existed in the region, the community's overt recollection turns on the notion of abolishment of slavery under Simcoe and NOTL as a terminus of the Underground Railroad (Shadd et al., 2002). This framing introduces a fleeting national shame for an *historic* injustice as a means to introduce the more robust narrative that emphasizes the move to eliminate slavery (albeit through two generations). Through its commemoration of the end of slavery, NOTL can claim for itself, and for Canada more broadly, an explicit engagement with past wrongs that produces a benevolent present: '*By witnessing what is shameful about the past, the nation can "live up to" the ideals that secure its identity or being in the present.*' (Ahmed, 2014, p. 109; original emphasis)

This specific recuperation of the past as the foundation for contemporary tourism is situated alongside a rejection of the modern in NOTL. As Urry and Larsen suggest, 'the quest for picturesque townscapes is fuelled by a widespread desire to travel back in time, to a supposedly Golden Age of romanticised escapism' (2011: 175). As Old Town has managed to avoid the construction of high density suburbs, multi-story parkades, and fast food, it has been able to claim a slowing of time 'to a more leisurely time' (Tourism Niagara-on-the-Lake, 2016) and so produce an aesthetic space that provides a seeming contrast to the compression of time and the materiality of space more often associated with modernity.

### *Food & wine: Farm-to-Table in NOTL*

In no small part, NOTL constructs itself as outside of modernity through its emphasis on locally produced food and wine. This emphasis is important to many tourists and many residents, though the link between heritage and wine – as a relatively recent new industry in the area, one that has replaced the historical production of tender fruits – is disputed by some residents (Ripmeester & Johnston, 2013). Both the Old Town and the Niagara Peninsula more broadly highlight food as a key element of the touristic experience (Telfer & Hashimoto, 2013). Due to its unique geological formation (Tinkler, 1994) and the buffering effects of Lake Erie, Lake Ontario, and the Niagara Escarpment, the Niagara peninsula has a relatively mild climate that situates it as part of Ontario's fruitbelt (Shaw, 1994). As such, the landscape is characterized by vineyards and orchards that produce cherries, peaches, plums, and nectarines – tender fruits that cannot be produced on a mass scale outside of the specific climatic requirements of NOTL and other small zones in Southern Ontario. The production of tender fruit in the area has given way to the emergence of wineries (Ripmeester & Johnston, 2013) and there are now more than 80 wineries in the region.

Given the area's agricultural capacity and its desire to produce itself as outside of the modern, consuming locally produced food and wine factors largely into regional tourism initiatives. This commitment to local food provides an alternative mode of tourism that reflects – at least superficially – a commitment to small-scale capitalism, in which tourism supports local agriculture directly and indirectly. These initiatives are evident through events such as the White Effect Dinner in June, which features 'catered picnics from Signature Kitchens[7] …, matching VQA wines from local wineries, local craft beer and terrific live music' (NOTL Official Visitor's Guide, 2015, p. 12) and 'Shaw Feature Lunches,' in which theatre events are linked with culinary offerings 'using the finest local & regional ingredients' (Prince of Wales Hotel and Spa, 2016).

Local food and wine alliances also exist (see also Telfer & Hashimoto, 2013). In addition to the numerous restaurants that incorporate multipage Niagara wine lists into their menus, networks are evident in initiatives such as LincFarm @ Southbrook, pork and lamb production at a winery. In its 2016 promotional pamphlet, LincFarm's meat is endorsed by the chefs at two of NOTL's well known restaurants. Wineries have also linked together to form loosely connected alliances such as Wine Country Vintners, a store front retailer and tasting boutique on the main street of NOTL that brings together wines from four local wineries. Similarly, Somewhereness is an affiliation of 12 wineries committed to 'small plots, sustainable practices, responsible stewardship and … winemaking [that] contributes to the elusive characteristics that flow through to your soul with every sip of our wine' (Somewhereness, 2018).

Upscale restaurants in NOTL include Treadwell (2018) and Backhouse (2018), which cite 'farm-to-table cuisine' and 'enlightened hospitality and cool climate cuisine,' respectively. These restaurants are locally owned and their menus highlight local food products with specific references to the farms that produced them. The menus at winery restaurants (see, for instance, Trius Winery at Hillebrand, 2018) highlight the use of local foods, while Ravine Vineyard Estate Winery (2018) goes to the extent of tending its own restaurant garden. These relationships between farms and wineries and the region's commitment to locally sourced food and crops are clearly mapped out – sometimes literally – for tourists. While the relationships between producers and retailers are explicit in many wineries and restaurants, the materiality of labour remains obscured.

## Global labour on local farms

In 1966, 264 Jamaican men traveled to Canada to provide their labour to tobacco farms in (Gabriel & MacDonald, 2011) southwest Ontario in two regions: Leamington and St. Catharines, the latter adjacent to NOTL (McLaughlin, 2009). Their travel was the outcome of an agreement between Canada and Jamaica– the Seasonal Agricultural Worker Program – that facilitated the movement of Jamaican farm labour into southwestern Ontario's tobacco belt. SAWP was intended to address farmers' claims that they could not secure adequate reliable workers during the six-week tobacco harvest. In addition to local labour, these farmers relied heavily upon domestic 'transient' labour, which included students and unemployed persons travelling from northern Ontario, Quebec, and eastern Canada, as well as tobacco workers coming to the area from the southern United States (Reid-Musson, 2014).

These domestic migrant workers possessed '"exit power" as a means of dispute resolution' (Reid-Musson, 2014, 170), which is to say that they were able to enact their mobility as a means of resistance to 'the conditions of seasonal, poorly remunerated, difficult and often dangerous farm work' (Reid-Musson, 2014, p. 162). With local workers, growers experienced a high degree of labour turnover. With the introduction of international labour, farmers were granted greater control over the behavior of workers both during and outside of working hours given their capacity to repatriate workers. As one farmer stated: 'I like Caribbean labor because I hold something over him. If he chatters too much or stays out too late at night I can send him home. You have no control over Canadian labor. You can't force them to stay' (Simcoe Reformer, 11 February 1977 cited in Reid-Musson, 2014, p. 168).

Within agriculture, poor working conditions persist. Agricultural work is deeply embodied work: not only does physicality inhere, but this physical labour takes place within an uncontrollable climate (excepting greenhouse work). Demands for labour are dictated by crops and by weather. Across Canada, these conditions of agriculture exempt it from many of the employment standards set out in provincial regulations. In Ontario, agricultural labourers are exempted from the sections of the *Employment Standards Act* that require overtime pay; public holidays or holiday pay; limits on daily and weekly hours of work; required daily eating periods; minimum rest periods between shifts; and mandatory weekly/bi-weekly rest periods (Ontario Ministry of Labour, 2018). Given the combination of the bio-physicality of agricultural labour and its exemption from many regulatory protections, it is unsurprising that farmers in Ontario (and across Canada) find it difficult to recruit local workers.

Until 2018, farmers had to demonstrate an inability to secure Canadian workers by posting an advertisement in the national Job Bank (or its provincial equivalent) as well as in local or regional newspapers, in local community centres or stores, or on employment websites for a minimum of two weeks to acquire an Labour Market Impact Assessment (LMIA). LMIAs are used to determine the impact of hiring a foreign worker on the Canadian job market. The application includes questions such as 'Will hiring a

TFW [temporary foreign worker] fill a labour shortage?' and 'Will the hiring of the TFW(s)… lead to job losses, or a reduction in the work hours… for Canadian/permanent resident employees in the employer's workforce or to the Canadian workforce more generally?' (Employment & Social Development Canada, 2018b). Notably, as of the spring of 2019, agricultural employers no longer have to provide evidence that they tried to recruit Canadians to work in order to qualify for an LMIA (Government of Canada, 2018), a policy shift that reflects the permanent shortcoming of domestic labour in the agriculture sector.

Officially a temporary work program, SAWP is the mainstay of employment for labour-intensive agriculture including tender fruits and grapes in Niagara, alongside the field vegetables and greenhouse crops concentrated in other parts of Ontario. For workers who participate in SAWP, the already high demands of agricultural labour are exacerbated by two specific terms of the agreement: 1) Workers' employment visas are closed, meaning they are tied to one employer;[8] 2) Workers may be repatriated for 'noncompliance, refusal to work, or any other sufficient reason' (Contract for the Employment in Canada of Commonwealth Caribbean Seasonal Agricultural Workers, 2018, p. 12). As a result, workers who want to continue participating in the program – in a given year, or over a series of years – are dependent on their specific employer identifying them as 'good' workers.

## Contradictions in NOTL: in/visibility

### *Visibility/invisibility*

Niagara-on-the-Lake is a centre of agriculture and of tourism, which, respectively, produce and consume landscape. As Urry and Larsen argue in their revision of Urry's seminal work on the tourist gaze, individuals direct their gaze towards 'features of landscape and townscape which separate them off from everyday experience' (2011: 4). In a space such as NOTL, which is decidedly and purposefully un-urban, rural life and landscapes become the recipient of tourists' gazes. The framing of NOTL as a heritage destination involves a combination of its townscape and its ruralness. In gazing upon the rural landscape, it is necessary for tourists to not only identify and frame certain elements that are designated as picturesque, but it also to actively un-see elements that disrupt the aesthetics of the countryside:

> a 'rural landscape' normally has erased from it farm machinery, labourers, tractors, telegraph wires, concrete farm buildings, motorways, derelict land, polluted water, nuclear power stations and dead and diseased animals. What people see is selective, and it is this focused gaze that is central to people's appropriation. The countryside is there to be gazed upon, and ideally one should not be gazing upon other people, whether workers or other tourists. (Urry & Larsen, 2011, p. 111)

Refining this observation, we argue that the presence of agricultural workers is particularly disruptive to this landscape. While the workers who provide direct service can be incorporated into the tourist gaze through the discourse of 'customer service,' the presence of those who labour in agricultural production contextualize the gastronomic tourist experience of NOTL within racialized physical labour. Thus, while fruit trees and blossoms and vineyards are picturesque, the individuals who spend up to 14 hours a day, frequently seven days a week, for up to eight months, are obscured. As in California, 'the popular image of the ... countryside erases the labor that creates the countryside. The blessed bounty of fertile fields, images of oranges and peaches hanging from trees, ignores the workers who maintain the landscape' (Wald, 2011, p. 574). Similarly, McLaughlin argues that in NOTL, migrant workers, although the mainstay of viticulture in the region, 'remain on the margins of society, largely invisible to wine consumers and to the public' (2013, p. 110).

In the fields that constitute the spaces of labour, it is through directed un/gazing that the workers who literally produce the landscape through their weeding, pruning, thinning, and harvesting are made invisible. Physically, the fields they tend are large and thus their time along the roads – points from which they are visible to tourists driving, being driven, or cycling through the area – is limited. Thus, the very nature of agricultural work facilitates a geographical separation of migrants from other Niagarans and tourists to the region and can heavily limit their visibility and the time they have to interact. During peak season (July and August) migrant workers will oftentimes be out in the fields from sunup to sundown, Monday to Saturday, and until late afternoons on Sundays.

At the same time that labour is rendered invisible, the conditions of work in the fields – the heat, the dust, the rain, the mud, the sprays (including from adjacent fields) – are discursively inconsistent with the temperate and controlled comfort expected of touristic experiences. This inconsistency is most explicit in the cancellation of one wineyard tour, in which one of the authors planned to participate for the specific purpose of viewing how the workers were viewed within the literal field of production. The temperature was over 30 degrees Celsius and the tour guide deemed the conditions too extreme to walk the rows of vines; instead, the participants participated in a wine and cheese pairing workshop in an airconditioned lounge with windows overlooking the fields in which the farm workers pruned vines for 12 hours.

To residents of NOTL, farm workers are more visible, though their numbers may be underestimated. In a conversation with a retail employee in Old Town to explain the purpose of being in NOTL, the employee suggested that there were about 100 workers in the area (in contrast to the approximately 3000 workers that arrive annually). In more isolated agricultural centres (e.g. Simcoe), workers are rendered (hyper)visible weekly as they arrive at shopping centres to buy groceries and other items, transfer money home, attend health clinics, and, time permitting, socialize with other workers in parking lots and nearby areas. In NOTL, however, workers are typically taken to St. Catharine's for their weekly errands as workers prefer the cheaper groceries and access to more retail services available outside of NOTL. To NOTL residents, farm workers may be visible at the Valu-Mart, the credit union, or other spaces in Virgil, a small town about 5 km from Old Town. The farm workers are also visible to locals as they ride bicycles – their primary form of independent transportation – between farms or for errands (see Reid-Musson, 2017). Yet because of they occupy spaces that are beyond the overtly tourist destinations (e.g. Virgil vs. Old Town), they are easily overlooked by tourists. Even when workers are rendered visible to local residents – for instance, when several buses converge to drop them off at grocery stores or when workers are seen riding bicycles – these residents may not understand the living and working conditions of those people who support their agricultural industry.

### *A palimpsest of labouring bodies*

In addition to the logistics of worker visibility, there is a willful invisibility about the workers working. In particular, the bodies of Black labourers working in the fields invites the spectre of slavery. While NOTL represents itself as the space in which Simcoe abolished slavery in Upper Canada in 1793, the fact that it is the space of slavery's abolishment identifies NOTL as a space in which slavery was practiced (see Power & Butler, 1993). Notably, Simcoe's legislation prohibited slavery only once the children of existing slaves reached the age of 25 and discouraged manumission in the interim. When it was passed, it did not free a single slave (Power & Butler, 1993). As such, it was only with the Imperial Act of 1833 that all slaves on British territories were freed (Riddell, 1932). The historical presence of slavery in Canada is echoed in the use of Chinese workers in the late 1800s to complete the Canadian Pacific Railway, the forced labour of Ukrainian enemy aliens during WWI to build infrastructure in Canada's National Parks, and the forced labour of Japanese Canadians interned during WWII for roadworks and on sugar beet farms. All of these events evidence the willingness of Canadians and the Canadian state to use non-whites (or, in the case of Ukrainians, not quite white enough) as a source of labour under conditions intolerable to those who are deemed sufficiently Canadian.

These historical practices sit uncomfortably alongside Canada's dominant national narrative of benevolence (especially vis-à-vis the United States). Formal acknowledgements, along with grants or reparations, have been made to Chinese Canadians (2006), Ukrainian Canadians (2008), and Japanese Canadians (1988). Though there have been calls to apologize and provide reparations for the slavery that was practiced in Nova Scotia before Canada became a country (Senate of Canada, 2018), there has been no response at the federal level. The lack of a response reflects national failure to engage the issue of historic practices of slavery. However, should such an apology be forthcoming, it should be received with great ambivalence.

As Ahmed (2000) has pointed out in her analysis of Australia's apology for its Terra Nullius policy (abandoned after the 1992 High Court Mabo decision), such public acts of shame and witnessing are not benign. They also serve to effectively situate exclusionary practices *in the past* and in so doing, create a definitive barrier between the overtly racist logics that underpinned the past and today's ostensibly tolerant multicultural states. Within such states, however, Hage argues that tolerant turns on not enacting intolerance, which means that the capacity for intolerance remains. As such, tolerant multiculturalism

> is a strategy aimed at reproducing and disguising relationships of power in society, or being reproduced through that disguise. It is a form of symbolic violence in which a mode of domination is presented as a form of egalitarianism. (2000: 86)

In this way, the racialized logics of the state are resilient and persist within Canada and, indeed, Niagara-on-the-Lake.

Jane Andres (2016), a local community member in NOTL, observes the narrative continuity between past practices of slavery in Palatine Hill and the contemporary experiences of the Jamaican migrant workers who are her 'neighbours,' albeit precariously so. She highlights the ways in which the rich farmland of NOTL was cleared by slaves 200 years ago; it is now tended by women and men whose ancestors were forcibly relocated through the slave trade. The presence of racialized, and especially Black, people participating in manual outdoor labour haunts NOTL (see Gordon, 1997). As Van Wagenen points out, 'an investigation of haunting is a history of the present. This is the nature of a haunting; it makes the absent past present in structuring the here and now' (Van Wagenen, 2004: 288). In Niagara, the history of slavery is recalled in the presence of globalized migrant labour. This palimpsest of bodies of colour labouring for the benefit of white bodies is inconsistent with Canadian narratives of multiculturalism and inclusion (Mackey, 1999), with the discourses of 'enchantment' (Thompson & Coskuner-Balli, 2007) and nostalgia (Roos et al., 2007) that collapse onto socially responsible consumption, and with the pleasure that informs tourist practices (Harrison, 2003; Urry & Larsen, 2011).

Even outside of their spatialized labour, migrant workers in the area are not integrated into the landscape of the peninsula. Their moorings – the bunkhouses in which they spend almost all of the time they are not working, because of a lack of safe mobility options (Reid-Musson, 2017) – are inconsistent with the aesthetics expected of quaint tourist spaces. Workers reside in small houses that range from adequate to derelict. Most these homes are not picturesque. Their porches, should they exist, are filled eclectic collections of used furniture. Doorsteps are ramshackle. Bicycles, used by the workers as their only autonomous means of transportation, are cluttered against trees or walls. Many of the yards feature large, productive gardens. However, these gardens are not situated within a manicured and tended setting. When the workers are home, the driveways host farm vehicles, some with cardboard over absent windows and others missing doors. While employers are required to provide washing and drying facilities (Employment & Social Development Canada, 2018c), a washer, we have been told anecdotally, may take the form of a bucket. Clotheslines are a common sight alongside farm workers houses and the workers' clothing reflects the workers' long and difficult labour. As such, their drying clothes are inconsistent with nostalgic views of the rural idyll.

### *Regulated invisibility*

While NOTL has a specific relationship to agricultural labour given its commitment to gastronomic tourism, the invisibility of agricultural labour is not specific to NOTL. Indeed, this absence is present across Canada due to the terms of SAWP. At the crux of the SAWP's role in (re)producing migrant invisibility is the clause that ties workers to the employers who requested them. With few exceptions[9], migrant workers are not permitted to seek employment opportunities at another farm once hired, and this leaves them completely dependent upon the farm owner. Workers understand that if they do anything deemed by their employer to be problematic, they will face one of two fates. In the first, the worker may not be 'named' again the next year and will thereby have their spot assigned to one of the many other individuals who want to participate in the program. In the second, the employer may terminate their employment and, with the assistance of the liaison, have the worker on a plane back to their home country in a matter of hours.

As noted above, the SAWP contract for both Mexican and Caribbean workers coming to Canada includes subjective language that enables farm owners to dismiss workers for vague and potentially pedantic reasons (Contract for the Employment in Canada of Commonwealth Caribbean Seasonal Agricultural Workers, 2018, p. 15). No further clarification is provided, and the resulting vagueness places the onus on workers to stay in their bosses' good graces, lest their employer perceive them as non-compliant. Under these conditions, the contestation of any on-the-job hazard, any labour law violation, or any other grievance forces workers to weigh the risk of their working conditions against the risk of losing their income. As one worker we spoke with put it, 'You have a choice [to speak out], but you have to know how to use that choice.' Given the potential outcomes of raising an issue with an employer, workers most often choose to remain silent, even when conditions are knowingly hazardous or in violation of other labour laws (Hennebry et al., 2016; McLaughlin, 2010).

Not only are workers reluctant to speak up about health and safety violations, they also reluctant to attend to their own health needs. Although migrant workers are legally entitled to provincial health care and employers are 'responsible for transportation to and from a hospital or clinic whenever the worker needs medical attention ... in a timely fashion' (Contract for the Employment in Canada of Commonwealth Caribbean Seasonal Agricultural Workers, 2018, p. 10), a worker who asks to be taken to hospital runs the risk of being seen as lazy or uncooperative (Hennebry et al., 2016; McLaughlin et al., 2014; Orkin et al., 2014; Preibisch & Otero, 2014).

Despite the formalization of worker protections in Ontario (and other provinces) through the *Occupational Health and Safety Act (*1990) and *Employment Standards Act (*2000; admittedly gutted by exemptions for agricultural workers), as well as through the bi-laterally negotiated contract, seasonal agricultural workers rely upon their employers to abide by these regulations. As noted above, Liaison Officers who are tasked with ensuring the program runs smoothly in Canada, are understood by workers to be allied with farmers and their government. Within an asymmetrical relationship – one in which Canada offers jobs to workers who have limited access to employment within their own countries and who come from countries that stand to benefit from the remittances provided by their outgoing workers – the smooth operation of the program favours Canada and its employers.

Migrant isolation and the subsequent dependence upon employers are exacerbated for migrants with limited English skills. The lack of comprehension of Spanish-speaking workers and some Patois speakers who are not confident in their English abilities serves to further the precarity of the workers. Basok (2002), Binford (2009), and Hennebry (2012) describe how employers use the threat of changing the countries from which they recruit, from English-speaking Jamaica to Spanish-speaking Mexico, as a means by which to instill fear, divide the workforce, and stifle employee dissent. Preibisch and Binford (2008, p. 16) call this practice 'country surfing,' explaining that employers engage in an ongoing 'quest for the most docile, exploitable labour force.'

Some employers have also expressed that having workers from English-speaking countries like Jamaica makes *them* feel vulnerable, in that these employers perceive there to be a greater risk that English-speaking workers have the capacity to complain about their working conditions or even walk away from the program to join the Canadian-Caribbean diaspora without official status (Preibisch & Binford, 2008).

Language barriers can also leave workers dependent upon employers for information regarding safety, health, and acute medical services. Through the SAWP as a whole, workers find themselves at an increased chance of on-site accidents, while unable to access healthcare services (Basok, 2002; Edmunds et al., 2011; McLaughlin, 2009; Narushima & Sanchez, 2014; Ramos et al., 2015). For those workers who do manage to see a physician or other healthcare specialist, language barriers and the limited number of on-demand translators in small towns and rural communities can hinder effective communication, compromising treatment with the possibility of future complications. The doctors who are available may not have been trained to be attentive to the problems for which agricultural workers might be uniquely at risk, including irritation and diseases of the skin, lungs and eyes (McLaughlin & Tew, 2018). Workers in small towns report feeling frequently dissatisfied with the care they do receive in this regard, leading them to self-medicate with drugs they bring from home (Díaz Mendiburo & McLaughlin, 2016; Hennebry et. al., 2015; Mysyk et al., 2009).

The (in)ability of migrant workers to access health services or language-training, or to participate in community events intended to include and support them (such as church services) is further restricted by their physical isolation. Given the rural context, there are frequently significant distances between farms and the towns that host resources useful to workers, as well as between farms themselves. Very few workers can afford their own vehicles and associated costs, although some are able to use the farms' vehicles (Reid-Musson, 2017)

NOTL is unique as a centre with a large number of migrant farm workers in which several community-drive initiatives have emerged to support workers. These initiatives include Bikes for Farmworkers, which sells safe bicycles to farm workers for $25 and provides repairs at cost (Bikes for Farmworkers, n.d.), and a free clothing 'store' that ran out of the garage of a community member until she moved; an alternate venue is currently being organized. NOTL also hosts a Niagara Migrant Workers Interest Group (2020) that seeks to provide social, legal, and health-related supports to workers. Quest Community Health Centre, based in St. Catharines, provides health clinics in English and Spanish approximately every two weeks in Virgil through the agricultural season (Quest Community Health Centre, 2017) through their Migrant Agricultural Worker Program. In 2017, a Peach Pickers' Picnic was organized by one community member to coincide with the Peach Festival (see opening epigraph). The Picnic revised its predecessor, the Workers' Welcome Concert that had run for the previous 11 years, as an explicit effort to recognize the labour of the workers and to render them visible: it is an event that draws hundreds of workers and thanks them with dinner and music alongside Niagara Stone Road, one of the busiest thoroughfares in the area. Despite these active efforts of community members to support the many farm workers in the region, the living and working conditions of many of the workers remain problematic and, at times, dangerous (Figure 3).

**Figure 3.** Peach Pickers Picnic. Since 2017 the Peach Pickers Picnic has coincided with the on-street dinner that is the concluding event of the region's Peach Festival. The community-organized celebration of migrant farm workers brings Caribbean and Mexican workers together with local community members for an evening of food and live music. © Kristin Lozanski

One source of significant risk to workers occurs beyond their living and working conditions, as they walk and ride bicycles as a means of independent travel. Because of its relatively small farms, workers in NOTL can get from their farm to other farms, as well as to nearby towns and to the city of St. Catharines by bicycle. Despite this relative autonomy, the long hours that workers are in the fields and packing houses (agricultural labourers in Ontario are not subject to a maximum number of working hours per day, nor mandatory breaks) mean that the opportunities for autonomous travel typically take place in the evenings or at night. After their shifts, workers are often exhausted and riding a bicycle is simply too much effort. Moreover, there have been a significant number of accidents and deaths involving migrant workers riding bicycles in rural Ontario. In a 2011 submission to Cycling Death Review undertaken by the Office of the Chief Coroner in Ontario, Justicia for Migrant Workers identify at least four farm workers who died while riding their bicycles between 2006 and 2011, while many others were seriously injured (Ramsaroop & Smith, 2011). In 2016, Andres Dominguez Moran was killed, and his co-worker Jesus Sanchez Feliciano sent to hospital with life-threatening injuries, after a car ran into them as they rode along the side of a rural highway (Noseworthy, 2016). Because of such accidents, some employers have begun to restrict their employees' access to bicycles (Reid-Musson, 2017), a move that both is protective and paternalistic and serves to reinforce employers' control over the mobilities of their workers. Given the lack of access to (safe) physical mobility, workers spend much of their non-working time stranded in their respective farmhouses.

Alongside the restricted mobilities of workers, many farm owners are reluctant to allow outside visitors. Fearing that allowing visitors to farm houses may distract workers from the actual farm labour, or the rest necessary to recuperate from their physical labour, some farm owners discourage social interaction by prohibiting outsider access without prior consent or even refusing to allow their workers to leave the farm with community volunteers to attend church services or public events. In this way, many workers are prevented from connecting with people including union workers, salespeople, researchers, journalists, and community members who may be able to provide important services and supports (Basok, 2002; Hennebry, 2012; Horgan & Liinamaa, 2017).

These multiple forms of isolation are embedded in the regulatory structure of SAWP. While the regulations to provide housing are presumably intended to support workers, the nature of that housing – under the surveillance (direct or otherwise) of the employers – exacerbates the social and geographic dis-integration of workers into local communities. This isolation, alongside the closed work permits and the 'non-compliance' clause with SAWP contracts render migrant agricultural workers invisible to most people moving through the spaces of their agricultural labour. The precarity of these workers and their invisibility are mutually reinforcing.

## Conclusion

The production of food and wine in Niagara-on-the-Lake is actively portrayed as local. In this way, the regions' gastrotourism industry implicitly recalls an idealized pre-modern era, in which direct relationships existed between food producers and food consumers. These imagined relationships contrast with those that dominate urban patterns of food consumption, which are often mediated by the transnational circulation and the corporate distribution (via large-scale grocery chains) of many food products. However, this narrative of local production and consumption obscures the ways in which food and wine production in NOTL is entirely dependent upon transnational labour.

A globalized labour force is essential as very few Canadians are willing to do agricultural work (and have been historically unwilling as well; Reid-Musson, 2014). As the bodies of agricultural labourers from Mexico and many parts of the Caribbean cross the Canadian border into NOTL's zone of ostensible locality, so too do globalized practices of exploitative labour. While Canada regulates labour and environmental conditions relatively rigorously, the workers who come to Canada through SAWP (and other migrants who come through the Temporary Foreign Worker program to participate in agriculture and other low-wage positions) constitute a site at which production costs may be managed. Workers tolerate minimum wage with no overtime or required breaks, violations of chemical safety practices (McLaughlin et al., 2014) and the possibility of immediate repatriation because there is a significant number of people in their home countries – and other 'competing countries' (Binford, 2009) – ready and willing to replace them. The willingness of workers to return to Canada year after year highlights the lack of opportunities in their home countries, including opportunities for relatively well paid labour, as well as opportunities for their children to become educated (according to the workers, education and associated costs absorb a significant portion of their remittances; see also Wells et al., 2014)[10].

However, this transnational circulation of labour is inconsistent with pre-modern systems of production and consumption. In this way, the idyll of local food and wine NOTL is dependent upon the obfuscation of not only agricultural labour in regional fields, but also the broader context of globalized, migrant labour: familial separation for up to eight months per year, poor living conditions for workers (see Schedule F; Employment & Social Development Canada, 2018c), and worker-reported non-compliance with mandated health and safety regulations (Sargeant & Tucker, 2009). Workers are able to send remittances to their families in their home countries only because they live in poverty, many residing in accommodations far inferior to that in their home countries, while they are in Canada. While many tourists wish to have an authentic experience of their destination (MacCannell, 1990), the actual 'backstage' of food and wine production in NOTL is disruptive to the discourse of ethics and personal connection that implicitly underpins the gastro-economy of the region. NOTL is not a place that has somehow managed to maintain pre-modern relationalities, but is instead structurally embedded in, and deeply dependent upon, the transnational exploitation associated with late capitalism.

## Notes

1. VQA is a quality designation of wine provided by a provincial body that regulates the labelling of wines and ensures that the wines are made from grapes grown and cellared in the province. (VQA 2016).
2. Caribbean signatories to the program include Jamaica, Barbados, Trinidad & Tobago, Anguilla, Antigua and Barbuda, Dominica, Grenada, Montserrat, St. Kitts-Nevis, St. Lucia, and St. Vincent and the Grenadines
3. Canada's Temporary Foreign Worker Program is a non-governmentally mediated program that brings agricultural workers from any country to stay and work in Canada on a two years work permit (Government of Canada, 2019)
4. Kayla Baumgartner was a Research Assistant for this project from April 2016 to August 2017 and visited NOTL four times to observe the townsite and to meet with workers.
5. While Niagara-on-the-Lake designates a region that includes the small towns of Virgil, Queenston, St. David's and the Old Town, as well as rural areas, in this paper, we focus on the Old Town because it contains the vast majority of the heritage designated properties (44 of 61; Niagara-on-the-Lake, 2014); it is also the site of the Shaw Festival theatres and the Niagara Historical Museum, and is the hub of tourist services, including bike rentals and wine tours.
6. Ultimately, this application of UNESCO World Heritage designation was unsuccessful (Zettel, 2017)
7. Signature Kitchens is comprised of 14 restaurants in NOTL 'commit[ted] to excellence through the use of high-quality local ingredients, a focus on outstanding VQA wines, and the delivery of professional service' (Signature Kitchens, 2016).
8. In Ontario, employees' work permits are tied to their employers. They may be transferred, with their permission, to another employer in the same sector. Transfers are also possible, though rare, in circumstances of abuse or exploitation (Contract… 2018). Many farm workers who labour in vineyards and orchards are transferred to either harvest apples or work in greenhouses in the fall.
9. Workers are permitted to change farms when it is agreed to by the worker and by both farm owners. In Ontario, this often happens in the fall, with workers typically transferring farms to pick apples. In 2014, remittances from Jamaican farm workers were reported to

be approximately 1.7B JMD/19.7M CAD (Smith-Edwards, 2015). Annual remittance flows from all countries account for approximately 15 percent of the Jamaica's gross domestic product (McIntosh, 2019).

10. In 2014, remittances from Jamaican farm workers were reported to be approximately 1.7B JMD/19.7M CAD (Smith-Edwards, 2015). Annual remittance flows from all countries account for approximately 15 percent of the Jamaica's gross domestic product (McIntosh, 2019).

## Acknowledgements

We are grateful to the men and women who support our agriculture and who have shared their time and stories generously.

## Disclosure statement

No potential conflict of interest was reported by the author(s).

## Funding

This research was supported by an Insight Development Grant 430-2014-00240 awarded by the Social Science and Humanities Research Council of Canada and research grants obtained from King's University College.

## References

Agriculture and Agri-Food Canada. (2017). An overview of the Canadian agriculture and agri-food system 2017. Retrieved October 29, 2018, from http://www.agr.gc.ca/eng/about-us/publications/economic-publications/an-overview-of-the-canadian-agriculture-and-agri-food-system-2017/?id=1510326669269#a4

Ahmed, S. (2000). *Strange encounters: Embodied others in post-coloniality*. Routledge.

Ahmed, S. (2014). *Willful subjects*. Duke University Press.

Andres, J. (2016). Lessons and legacies in our backyard [Unpublished].

Backhouse. (2018). Backhouse. Retrieved June 7, 2018, from backhouse.xyz

Basok, T. (2002). *Tortillas and tomatoes: Transmigrant Mexican harvesters in Canada*. McGill-Queen's University Press.

Basok, T. (2003). Mexican seasonal migration to Canada and development: A community-based comparison. *International Migration*, *41*(2), 3–26. https://doi.org/10.1111/1468-2435.00233

Basok, T. (2004). Post-national citizenship, social exclusion and migrants rights: Mexican seasonal workers in Canada. *Citizenship Studies*, *8*(1), 47–64. https://doi.org/10.1080/1362102042000178409

Baxter-Moore, N., & Charest, C. (2013). Constructing authenticity: Architecture and landscape in Niagara's wineries. In M. Ripmeester, P. G. Mackintosh, & C. Fullerton (Eds.), *The world of Niagara wine* (pp. 259–279). Wilfred Laurier University Press.

Bikes for Farmworkers. (n.d.) Poster. Retrieved Feb 24, 2020, from https://notl.civicweb.net/document/16156

Binford, L. (2009). From fields of power to fields of sweat: The dual process of constructing temporary migrant labour in Mexico and Canada. *Third World Quarterly*, *30*(3), 503–517. https://doi.org/10.1080/01436590902742297

Binford, L. (2013). *Tomorrow we're all going to the harvest: Temporary foreign worker programs and neoliberal political economy*. University of Texas Press.

Bisby, A. (2019, July 4). GO Train fuels Niagara housing boom. The National Post. Retrieved 14 Jan, 2019 from https://nationalpost.com/life/homes/go-train-fuels-niagara-housing-boom

Contract for the Employment in Canada of Commonwealth Caribbean Seasonal Agricultural Workers. (2018). Retrieved October 1, 2018, from https://www.canada.ca/content/dam/canada/employment-social-development/migration/documents/assets/portfolio/docs/en/foreign_workers/hire/seasonal_agricultural/documents/sawpcc2018.pdf

Crag and Canyon. (2007, May 8). Other towns don't ban chain stores. The Crag and Canyon. Retrieved November 8, 2016, from http://www.thecragandcanyon.ca/2007/05/08/other-towns-dont-ban-chain-stores

Díaz Mendiburo, A., & McLaughlin, J. (2016). Vulnerabilidad estructural y salud en los trabajadores agrícolas temporales en Canadá. *Alteridades*, *26*(51), 85–95.

Dubinsky, K. (1999). *The second great disappointment: Honeymooning and tourism at Niagara Falls [electronic resource]*. Between the Lines. Accessed via Scholars Portal Canadian Publishers Collection. 8 November 2016.

Edmunds, K., Berman, H., Basok, T., Ford-Gilboe, M., & Forchuk, C. (2011). The health of women temporary agricultural workers in Canada: A critical review of the literature. *The Canadian Journal of Nursing Research = Revue Canadienne de Recherche en Sciences Infirmieres*, *43*(4), 68–91.

Employment and Social Development Canada. (2018a, May 3). Temporary foreign workers program operations: Overview and efforts to improve processing. Canadian Federation of Agriculture Ministerial Roundtable. Retrieved January 7, 2019 from https://www.cfa-fca.ca/wp-content/uploads/2018/06/ESDC_ProcessingOverview.pdf

Employment and Social Development Canada. (2018b). Labour market impact assessment application – Seasonal Agricultural Worker Program. Retrieved October 29, 2018, from https://catalogue.servicecanada.gc.ca/content/EForms/en/CallForm.html?Lang=en&PDF=ESDC-EMP5389.pdf

Employment and Social Development Canada. (2018c). Schedule F – Housing inspection report – Seasonal Agricultural Worker Program and agricultural stream. Retrieved October 29, 2018, from https://catalogue.servicecanada.gc.ca/content/EForms/en/CallForm.html?Lang=en&PDF=ESDC-EMP5598.pdf

Employment Standards Act. (2000) Ontario. Retrieved January 21, 2019, from https://www.ontario.ca/laws/statute/00e41

Gabriel, C., & MacDonald, L. (2011). Citizenship at the margins: The Canadian Seasonal Agricultural Worker Program and civil society advocacy. *Politics & Policy*, *39*(1), 45–67. https://doi.org/10.1111/j.1747-1346.2010.00282.x

Gordon, A. (1997). *Ghostly matters: Haunting and the sociological imagination*. University of Minnesota Press.

Government of Canada. (2018, September 18). Hire a temporary worker through the Seasonal Agricultural Worker Program – Recruitment and advertisement. Retrieved October 29, 2018, from https://www.canada.ca/en/employment-social-development/services/foreign-workers/agricultural/seasonal-agricultural/recruitment.html

Government of Canada. (2019, May 2). Hire a temporary foreign worker through the Agricultural Stream – Next steps. Retrieved Feb 24, 2020, from https://www.canada.ca/en/employment-social-development/services/foreign-workers/agricultural/agricultural/after.html

Graham, B., Ashworth, G. J., & Tunbridge, J. E. (2000). *A geography of heritage: Power, culture and economy*. Oxford University Press.

Harrison, J. D. (2003). *Being a tourist: Finding meaning in pleasure travel*. UBC Press.

Healy, R. G. (2006). The commons problem and Canada's Niagara Falls. *Annals of Tourism Research*, *33*(2), 525–544. https://doi.org/10.1016/j.annals.2006.01.003

Hennebry, J. (2012). Permanently temporary? Agricultural migrant workers and their integration in Canada. *IRPP Study*, *1*(26), 1.

Hennebry, J. (2014). Transnational precarity: Women's migration work and Mexican seasonal agricultural migration to Canada. *International Journal of Sociology*, *44*(3), 42–59. https://doi.org/10.2753/IJS0020-7659440303

Hennebry, J., McLaughlin, J., & Preibisch, K. (2016). Out of the loop: (In)access to health care for migrant workers in Canada. *Journal of International Migration and Integration*, *17*(2), 521–538. https://doi.org/10.1007/s12134-015-0417-1

Henry, N. (2010). *Emancipation day: Celebrating freedom in Canada*. Natural Heritage Books.

Horgan, M., & Liinamaa, S. (2017). The social quarantining of migrant labour: Everyday effects of temporary foreign worker regulation in Canada. *Journal of Ethnic and Migration Studies*, *43*(5), 713–730. https://doi.org/10.1080/1369183X.2016.1202752

MacCannell, D. (1990). *The tourist: A new theory of the leisure class*. University of California Press.

Mackey, E. (1999). *The House of Difference: Cultural Politics and National Identity in Canada*. Routledge.

Mayell, S. (2016). *Up-rooted lives, deep-rooted memories: Stress and resilience among Jamaican agricultural workers in Southern Ontario* [MA Thesis]. McMaster University.

McLaughlin, J. (2009). *Trouble in our fields: Health and human rights among Mexican and Caribbean migrant farm workers in Canada* [PhD Dissertation].

McLaughlin, J. (2010). Classifying the 'ideal migrant worker': Mexican and Jamaican transnational farmworkers in Canada. *Focaal*, *2010*(57), 79–94. https://doi.org/10.3167/fcl.2010.570106

McLaughlin, J. (2013). The hands behind the harvest: Migrant workers in Niagara's wine industry. In M. Ripmeester, P. G. Mackintosh & C. Fullerton (Eds.), *The world of Niagara wine*, (pp. 109–124). Wilfred Laurier University Press.

McLaughlin, J., & Tew, M. (2018). Migrant farm worker health: Unique strategies for a unique population. In A. N. Arya & T. Piggott (Eds.), *Under-served: Health determinants of Indigenous, inner-city, and migrant populations in Canada* (pp. 253–264). Canadian Scholars Press.

McLaughlin, J., Hennebry, J., & Haines, T. (2014). Paper versus practice: Occupational health and safety protections and realities for temporary foreign agricultural workers in Ontario. *Perspectives Interdisciplinary Journal of Work and Health*, *16*(2), 2–17. https://doi.org/10.4000/pistes.3844

McIntosh, D. (2019). Migration has helped shape Jamaica's economic and sociocultural frameworks. *Jamaica Information Service*. Retrieved February 25, 2020, from https://jis.gov.jm/migration-has-helped-shape-jamaicas-economic-and-sociocultural-frameworks/

Morgan, C. (2015). *Creating colonial pasts: History and commemoration in Southern Ontario, 1860–1980*. University of Toronto Press.

Mysyk, A., England, M., & Gallegos, J. A. A. (2009). A case for certified interpreters for participants in the Canada/Mexico Seasonal Agricultural Workers Program. *Human Organization*, *68*(3), 318–327. https://doi.org/10.17730/humo.68.3.6g012756050r04h8

Narushima, M., & Sanchez, A. L. (2014). Employers' paradoxical views about temporary foreign migrant workers' health: a qualitative study in rural farms in southern Ontario. *International Journal for Equity in Health*, *13*(1), 65. https://doi.org/10.1186/s12939-014-0065-7

Niagara Falls Review. (2015, September 23). Army of migrant workers power Niagara's farms. *Niagara Falls Review*. Retrieved June 7, 2018, from https://www.niagarafallsreview.ca/news-story/8182359-army-of-migrant-workers-power-niagara-s-farms/

Niagara Migrant Workers Interest Group. (2020). About. Retrieved February 24, 2020, from https://www.nmwig.ca/site/about-niagara-migrant-workers-interest-group

Niagara Parks. (2020). Corporate Governance. Retrieved February 24, 2020, from https://www.niagaraparks.com/corporate/about-us/corporate-governance/

Niagara-on-the-Lake. (2014). Official plan review: Heritage, arts and culture. Retrieved November 8, 2016, from https://notl.civicweb.net/document/5053/Official%20Plan%20Review%20-%20Heritage,%20Arts,%20and%20Culture.pdf?handle=164E814290B24D3BB0E1BB39B1A762AD

Niagara-on-the-Lake Application Submission. (2017). Retrieved February 24, 2020 from https://notl.civicweb.net/document/7366

Noseworthy, K. (2016, August 3). It's a great loss' says farmer of migrant worker's death. *The Hamilton Spectator.* Retrieved Jan 14, 2019, from https://www.thespec.com/news-story/6794578–it-s-a-great-loss-says-farmer-of-migrant-worker-s-death/

NOTL Official Visitor's Guide. (2015–16). The white effect dinner.

Occupational Health and Safety Act. (1990) Ontario. Retrieved January 21, 2019, from https://www.ontario.ca/laws/statute/90o01

Ontario Ministry of Labour. (2018). Industries and jobs with exemptions or special rules: Agriculture, growing, breeding, keeping and fishing. Retrieved October 1, 2018 from https://www.ontario.ca/document/industries-and-jobs-exemptions-or-special-rules/agriculture-growing-breeding-keeping-and-fishing#section-6

Orkin, A. M., Lay, M., McLaughlin, J., Schwandt, M., & Cole, D. (2014). Medical repatriation of migrant farm workers in Ontario: A descriptive analysis. *CMAJ Open*, *2*(3), E192–198. https://doi.org/10.9778/cmajo.20140014

Power, M., & Butler, N. (1993). *Slavery and freedom in Niagara*. Niagara Historical Society.

Preibisch, K. (2004). Migrant agricultural workers and processes of social inclusion in Rural Canada: Encuentros and desencuentros. *Canadian Journal of Latin American and Caribbean Studies*, *29*(57-58), 203–239. https://doi.org/10.1080/08263663.2004.10816857

Preibisch, K. (2010). Pick-your-own labor: Migrant workers and flexibility in Canadian agriculture. *International Migration Review*, *44*(2), 404–441. https://doi.org/10.1111/j.1747-7379.2010.00811.x

Preibisch, K., & Binford, L. (2008). Interrogating racialized global labour supply: An exploration of the racial/national replacement of foreign agricultural workers in Canada. *Canadian Review of Sociology/Revue Canadienne de Sociologie*, *44*(1), 5–36. https://doi.org/10.1111/j.1755-618X.2007.tb01146.x

Preibisch, K., & Otero, G. (2014). Does citizenship status matter in Canadian agriculture? Workplace health and safety for migrant and immigrant laborers. *Rural Sociology*, *79*(2), 174–199. https://doi.org/10.1111/ruso.12043

Prince of Wales Hotel & Spa. (2016, July 11). 'The adventures of the black girl in her search for god' Shaw Feature Lunch. *Pamphlet distributed at Shaw Theatre performance.*

Quest Community Health Centre. (2017). MAWP. Retrieved February 24, 2020, from http://questchc.ca/participate-at-quest/mawp/

Ramos, A. K., Su, D., Lander, L., & Rivera, R. (2015). Stress factors contributing to depression among Latino migrant farmworkers in Nebraska. *Journal of Immigrant and Minority Health*, *17*(6), 1627–1634. https://doi.org/10.1007/s10903-015-0201-5

Ramsaroop, C., Smith, A. (2011, November 28). Re: Bicycle deaths review. Retrieved January 14, 2019, from http://www.justicia4migrantworkers.org/wp-content/uploads/2014/06/Bike-Deaths-Review-Submission-2011.pdf

Ravine Vineyard Estate Winery. (2018) Farm Dining. Retrieved June 7, 2018, from https://ravinevineyard.com/farm-dining#!restaurant

Reid-Musson, E. (2014). Historicizing precarity: A labour geography of 'transient' migrant workers in Ontario tobacco. *Geoforum*, *56*, 161–171. https://doi.org/10.1016/j.geoforum.2014.07.005

Reid-Musson, E. (2017). Grown close to home™: Migrant farmworker (im)mobilities and unfreedom on Canadian family farms. *Annals of the American Association of Geographers*, *107*(3), 716–730. https://doi.org/10.1080/24694452.2016.1261683

Riddell, W. R. (1932). Additional Notes on Slavery. *The Journal of Negro History*, *17*(3), 368–377. https://doi.org/10.2307/2714282

Ripmeester, M., & Johnston, R. (2013). New wine in old wineskins: Marketing wine as agricultural heritage. In M. Ripmeester, P. G. Mackintosh & C. Fullerton (Eds.), *The world of Niagara wine* (pp. 249–258). Wilfred Laurier University Press.

Roger, K., & Mignone, J. (2018). Living your ethics: "It's" not just a dusty document. In S. Kleinknecht, L.J. van den Scott, & C. B. Sanders (Eds.), *The craft of qualitative research: A handbook* (pp. 46-52). Canadian Scholars Press.

Roos, G., Terragni, L., & Torjusen, H. (2007). The local in the global: Creating ethical relations between producers and consumers. *Anthropology of Food*, *2*(S2), 1-11. https://doi.org/10.4000/aof.489

Sargeant, M., & Tucker, E. (2009). Layers of vulnerability in occupational safety and health for migrant workers: Case studies from Canada and the UK. *Policy and Practice in Health and Safety*, *7*(2), 51–73. https://doi.org/10.1080/14774003.2009.11667734

Senate of Canada. (2018). Black history experts call for apology, reparations. Retrieved January 7, 2019, from https://sencanada.ca/en/sencaplus/news/black-history-experts-call-for-apology-reparations/.

Shadd, A., Cooper, A., & Smardz Frost, K. (2002). *The underground railroad: Next stop*. Dundurn Press.

Shaw, T. B. (1994). Climate of the Niagara region. In H.J. Gayler (Ed.), *Niagara's changing landscape* (pp. 111–138). Carleton University Press.

Signature Kitchens. (2016). The culinary capital of Canada. Retrieved Nov 14, 2016, from http://www.signaturekitchensofniagaraonthelake.com/

Smith-Edwards, A. (2015, January 6). Jamaica-Canada employment programme pumps $1.7 billion in economy. *Jamaica Information Service*. Retrieved March 10, 2020, from https://jis.gov.jm/jamaica-canada-employment-programme-pumps-1-7-billion-economy/

Somewhereness. (2018). Origins. Retrieved June 7, 2018m from www.somewhereness.com

Telfer, D. J., & Hashimoto, A. (2013). Wine and Culinary Tourism in Niagara. In M. Ripmeester, P. G. Mackintosh & C. Fullerton (Eds.), *The world of Niagara wine* (pp. 281–299). Wilfred Laurier University Press.

Thompson, C. J., & Coskuner-Balli, G. (2007). Enchanting ethical consumerism: The case of community supported agriculture. *Journal of Consumer Culture*, *7*(3), 275–303. https://doi.org/10.1177/1469540507081631

Tinkler, K. J. (1994). Entre lacs: A postglacial peninsula physiography. In H. J. Gayler (Ed.), *Niagara's changing landscape* (pp.13–51). Carleton University Press.

Tourism Niagara-on-the-Lake. (2016). About Niagara-on-the-Lake. Retrieved November 8, 2016, from https://www.niagaraonthelake.com/page/About_Niagara-on-the-Lake

Town of Niagara-on-the-Lake. (2013). NOTL official plan 2013 office consolidation. Retrieved November 8, 2016, from https://notl.civicweb.net/document/5401

Treadwell. (2018) About us. Retrieved June 7, 2018, from treadwellcuisine.com

Trius Winery at Hillebrand. (2018). Trius winery restaurant. Retrieved June 7, 2018, from http://www.triuswines.com/

Urry, J. (1990). *The tourist gaze*. Sage.

Urry, J., & Larsen, J. (2011). *The tourist gaze 3.0*. Sage.

Van Wagenen, A. (2004). An epistemology of haunting: A review essay. *Critical Sociology*, *30*(2), 287–298. https://doi.org/10.1163/156916304323072116

Voices of Freedom. (2018). Home. Retrieved October 30, 2018.

VQA. (2016) VQA Act & Regulations. Retrieved November 14, 2016, from http://www.vqaontario.ca/Regulations/Standards

Wald, S. D. (2011). Visible farmers/invisible workers: Locating immigrant labor in food studies. *Food, Culture & Society*, *14*(4), 567–586. https://doi.org/10.2752/175174411X13046092851479

Wells, D., McLaughlin, J., Lyn, A., & Mendiburo, A. D. (2014). Sustaining precarious transnational families: The significance of remittances from Canada's Seasonal Agricultural Worker Program. *Just Labour*, *22*, 144–166. https://doi.org/10.25071/1705-1436.9

Zettel, M. (2017, December 22). NOTL left off Canada's updated UNESCO tentative list of heritage sites. *Niagarathisweek.com*. Retrieved February 24, 2020.

# Q method finds anti-refugee sentiments on Yemeni migration to Jeju

Armigon Akhmedov, William Cannon Hunter and Jeong-Gil Choi

**ABSTRACT**

Tourism and migration are conceptually linked in terms of the geographical movement of people. In 2018 a large group of Yemeni refugee migrants arrived in Jeju Island, South Korea because of an unforeseen confluence of political and economic pressures in Yemen and a combination of a visa-free policy and cheap airfare from Malaysia to Jeju. They are now semi-permanent residents on the island with no option to re-migrate. A multidimensional set of theoretical themes including social exchange and role identity among stakeholders; power and alienation; and trust and social responsibility are used to evaluate residents' subjectivities toward Yemeni refugee migrants to Jeju. This is applied to the theoretical interpretation of the public discourse on Yemeni migration to identify residents' attitudes using Q method. Four clusters of subjectivity are found, grouped around an anti-refugee sentiment. Interaction with and responsibility for this refugee immigrant population is largely rejected. Practical implications are that structural changes in policy or attitude changes through education are necessary to secure any form of social sustainability for the resident-refugee relationship in Jeju. The implication is that the set of theoretical themes used in this study offers a valuable path to finding empirical solutions to the refugee migration problem in tourism research.

**摘要**

就人类的地理流动来说，旅游和迁移在概念上是有联系的。2018年，大批也门避难者移民抵达韩国济州岛，原因是也门的政治和经济压力、济州岛的免签政策和从马来西亚到济州岛的廉价机票，这些因素出人意料地交织在一起。他们现在是岛上的半永久居民，没有重新迁移的想法。本研究评估了当地居民对前来济州岛的也门难民移民的主观感受，涉及社会交换和利益相关者之间的角色认同、权力和排斥以及信任和社会责任多个理论主题。本研究使用Q方法识别居民的态度，将此应用于也门移民公共话语的理论解释中。本研究发现了一个围绕反难民情绪分组的四个主观性看法的集群。与这些难民移民人口的互动及责任在很大程度上被证实。本研究的实践启示是通过教育实现本地政策或居民态度的结构性改变，对确保济州岛居民-难民关系的任何形式的社会可持续性都是有必要的。这意味着，本研究所使用的理论主题为旅游研究中难民移民问题的解决经验提供了一条有价值的路径。

## Introduction

The migration-tourism link has broadened the horizons of tourism geography researchers and deepened the conceptualization of global mobility across national boundaries (Boyle et al., 2014). It has also complicated traditional thinking on tourism as a driver for destination economy and a contributing sector to quality of life (Andereck et al., 2005; Lusticky & Bednarova, 2018). Refugee migration in particular profoundly affects the destination (Williams & Hall, 2000a). As residents begin to recognize that they are involved as stakeholders in destination tourism they will express strong views on the effects of such movement, especially when it does not meet desired economic results and social outcomes. If balanced and meaningful relationships in a social movement are not realized (Griffiths & Sharpley, 2012), then the expectations in the relationship between the migrant-tourist link and the host relationships may be undermined (Gursoy et al., 2010; Prayag et al., 2013). This threat is particularly acute when residents bear the cost of the sudden arrival of large refugee migrant groups.

This paper explores the migration-tourism link in the context of refugee migration and its potential effects on tourism. The research focus is on residents' subjectivities on tourism and Yemeni migration in Jeju Island, South Korea. Beginning in 2002, a visa-free entry policy for Malaysia was implemented to offset declining tourist arrivals from China. Cheap direct flights between Kuala Lumpur, Malaysia and Jeju were also promoted. This rather normal marketing scheme, however, led unexpectedly to the sudden arrival of a large group of Yemeni refugee migrants via Kuala Lumpur in 2018. Without the option to re-migrate (Williams et al., 2018), this group of more than 550 refugees has caused a crisis in Jeju Island. In the media, resident stakeholder attitudes towards the Yemenis seems mixed, ranging from sympathy to indifference to opposition.

Residents' subjectivity, in this sense, is a discourse that is not fully known to any single individual. It is based on facts, anecdotes, and experiences that may not be entirely accurate or true, but it does work to form the 'talk' and 'deeds' of the event (Hollinshead, 2007, p. 293). Public reaction to these newly arrived refugees deserves a closer look, to find out how residents react to a narrative, based in theory, on the effects of this dimension of the migrant-tourism phenomenon (Pêcheux, 1982). This inspection is of particular interest in light of the highly contested effects, or impacts, or refugee migration on destination society and economy (Dibeh et al., 2019).

In this paper, the migration-tourism link concerning refugees is explored using a set of theoretical themes, including social exchange and role identity among stakeholders; power and alienation; and trust and social responsibility. The purpose of this study is to apply this multidimensional set of themes to the interpretation of the public discourse in Jeju and to engage a sample of resident-stakeholders in forging their held views on Yemeni refugee migration (Williams et al., 2000). To this effect, this study employs Q method (Stephenson, 1993) as a means to grasp the range of held or operant subjectivities at work within this public discourse. This approach works to anticipate community support or opposition to refugee migrants and their associated social and economic issues. The benefits of identifying resident subjectivities include more informed and effective decision making in forging tourism development plans or policies and more effective practice in strategically undoing the effects of haphazard construction of infrastructure or the promulgation of distorted representations of the community, its member demographics, or views on the destination itself (Hunter, 2013). To this effect, the research goals in this study are as follows:

- To position the migration-tourism link in the context of three theoretical themes including social exchange and role identity among stakeholders, power and alienation, and trust and social responsibility:
- To extract key features of the three theoretical themes as they apply to the Yemeni refugee migrant crisis in Jeju Island:
- To employ a Q method research approach to identify residents' operant subjectivities toward Yemeni refugee migrants in Jeju:
- To identify and interpret the theoretical connection between residents' attitudes toward refugees and the migration-tourism link:
- To offer practical and academic implications concerning Yemeni immigration on Jeju tourism and the Korean nationalistic outlook.

## Literature review

The migration-tourism link represents a widened conceptualization of global mobility that frames movement across national boundaries (Boyle et al., 2014). In this context, the traditional notion of travel as a short-term visit has become intertwined with other events of 'varying duration' (Williams & Hall, 2000a, p. 2). These events can be described as *voluntary migration* and *refugee migration*. The difference between these two kinds of events is distinguished by motivation. The voluntary migrant chooses mobility in order to pursue new opportunities and experiences (pull motivation) whereas the refugee migrant has been forced to escape their country of residence for social, political, economic or environmental reasons (push motivation) (Ivanov & Stavrinoudis, 2018). Migration in both forms impacts destination countries in a variety of ways. It can enrich the destination culture, enhance the tourism product and widen the labor pool. But it can also cause wage deflation, a reduction in tax revenues and create social tension (UNWTO, 2010; Moufakkir, 2014). The complexity of the migration-tourism link in terms of forms and impacts is of growing concern for tourism researchers (Illés & Michalkó, 2008).

Voluntary migration refers to global mobility by choice. Wealthy retirees and pre-retirees are highly sought after because they are perceived to be financially independent with the potential to contribute to the destination economy (Hudson et al., 2019). These 'lifestyle migrants' (Montezuma & McGarrigle, 2019, p. 218) are equipped for residential investment. Voluntary migrants include a variety of 'long term travelers'. They are second-home-owners, exchange students or labor migrants and they tend to re-direct the flow of travel by occasionally re-migrating to their countries of origin to visit friends and relatives (Filimonau & Mika, 2019). In this sense, they have adopted a transnational identity where the meaning of 'home' has become blurred (Ndione et al., 2018). Those who voluntarily migrate for tourism employment and other workplace experiences influence chain migration and recruitment practices. They tend to access multiple social networks that improve their cultural experience (Janta et al., 2011).

In the migration-tourism mobility link, refugee migration generally refers to global mobility for the purpose of seeking asylum. The refugee is economically disadvantaged and travel is one-way and relatively permanent. Refugee migration is characterized as a crisis rather than an opportunity. In voluntary migration, individuals or small groups of people re-locate to other countries and maintain the option to re-migrate (Williams et al., 2018).

The Refugee Convention states that a person can be qualified as a refugee if that person has previously been considered a refugee under prior treaty arrangements. In addition, if that person is outside of the country of their nationality (or without nationality) and not able or willing to seek the protection that country due to a substantiated fear of being persecuted for reasons of race, religion, nationality, membership in a particular social group or political opinion (Worster, 2012), they are a refugee. Therefore, a refugee may not share all the same characteristics as a refugee migrant or others involved in other forms of mobility related to the migrant-tourism link.

In the context of this paper, Yemeni refugee migrants arriving in Jeju face a crisis shaped by several factors: they arrived relatively suddenly and in large numbers; they are perceived as an ethnically or religiously homogeneous group; they may have difficulty assimilating into destination culture and economy; their ethnic or religious identities may be perceived to be conflict with local norms; and theoretically they may cause economic and social crises at the destination resulting in a decline in tourist visitation and damage to the destination image (Ivanov & Stavrinoudis, 2018). True or not, the potential benefits of refugee migration remain highly contested (Dibeh et al., 2019). In this paper, the social context, rather than any 'true' definition of Yemeni migration in Jeju Island is explored from the view of Jeju residents as expressed as forms of operant subjectivity.

## *Jeju island and Yemeni migrants*

Jeju is a key international and domestic tourism destination in South Korea and has been the site for testing new tourism development projects and strategies including exclusive visa-waiver programs for a number of markets to strengthen the destination image and increase international visits. One of these programs, visa-free entry for Malaysia beginning in 2002, was implemented to offset a sudden drop in Chinese tourists due to political reciprocity concerning Korea's intentions to deploy the Terminal High Altitude Area Defense (THAAD) system. In addition, direct flights were opened between Jeju island and Kuala Lumpur, Malaysia four times a week by Air Asia, which offered ticket deals as cheap as USD$70. The promotion, intended to boost Malaysian tourist visits unintentionally allowed the entry of more than 550 Yemeni refugees seeking asylum (Casado-Díaz et al., 2014). The refugees, fleeing civil war in Yemen by way of Malaysia, had outstayed their free three-month Malaysian visa status and moved on to Jeju where they have stayed. Refugees began arriving in Jeju in 2016 followed by a rapid increase in arrivals by early 2018 (No Cut News, 2018). For refugees, there is no return trip and their presence has raised concerns among residents

(Williams & Hall, 2000b). The motivation behind the arrival of these refugee migrants can be described as involving these various home country stressors (push factors) and policy/travel channels (structural factors).

In previous literature, anti-refugee and particularly anti-Muslim sentiments have been seen to be driven by fears that refugee resettlement threatens a destination's 'social, economic and moral order' (Nagel, 2016, p. 283). Even Muslim tourists and voluntary migrants may experience a cold reception at the destination where they are perceived to be religious or ethnic 'others' who represent a threat (Stephenson & Ali, 2010). In the context of European Islam (Allen, 2007) and in the media, Muslim refugees have been colored by basic Islamophobic sentiments (Gardner et al., 2008), exacerbated by 9/11 and the emergence of ISIL and global terror attacks.

No such attacks have taken place in Korea. Rather, a unique view on Muslims exists, partially colored by global media (Jeon, 2011) but mixed with a specifically Korean perspective. Korean ethnonationalism fosters a homogenized 'Koreanness' based on a so-called 'pure' blood nation (Jamass, 2014) that works as a foundational distrust of any 'Other'. In Korea, anti-Muslim sentiment is based on a perception of Islam as a foreign religion rather than a feature of growing multiculturalism in society (Sheikh, 2019a). This prejudice is directed at new migrants as well as native-born Korean-Muslim converts. But paradoxically Korean tourism policy has targeted Muslim visitors through the Hallyu (Korean Wave) product and the Halal policy (Eum, 2017). The actual view of Jeju residents on Yemeni migrants has not been fully explored as a social discourse.

Recent research has explored the general opinions circulating on media channels, through government leaders and resident stakeholders, expressing varying attitudes toward the new Yemeni refugees ranging from hostility to ambivalence to sympathy and acceptance (Kwon, 2019; Sheikh, 2019b). Residents might hold high expectations for the economic contribution of migration while harboring concerns regarding migrant cultural and religious differences (El Aouni et al., 2013). Forced migration does not necessarily have to have a detrimental effect on the local tourism economy in spite of public opinion to the contrary (Dibeh et al., 2019). In fact, public opinion may not be universally opposed to the arrival of refugee groups. In some cases, local stakeholders may feel threatened by such new arrivals while NGO members or volunteers or others in official positions might support victims of forced migration (Tsartas et al., 2019).

The predicament of Yemeni refugee migrants in Jeju and residents' attitudes towards them can be explored through three theoretical themes based on the migration-tourism link (Boyle et al., 2014). First, social exchange and role identity among stakeholders describe how refugees offer an opportunity to generate wealth and social value in Jeju (Andereck et al., 2005). Second, power and alienation refer to how the asymmetrical power position (Hall, 1994) of the refugee versus the resident precipitates hostility when altruistic intentions are eclipsed by the pursuit of short term benefits (Pêcheux, 1982). Third, trust and social responsibility reveal how residents could extend goodwill to refugees (Hunter, 2001) as they in turn act to sustain the good opinion of their hosts in the interest of group pride and identity (Pettit, 1995). These three theoretical themes are described in the following sections.

### *Social exchange and role identity among stakeholders*

Social exchange theory is concerned with interactions between two parties (Cook et al., 2013) for seeking material, social or psychological value (Andereck et al., 2005). In the context of this paper, social exchange is contingent upon role identities that are adopted by national residents (as tourism stakeholders) or ascribed to the 'Other' (as refugee migrants). Since social exchange is driven by the generation of wealth, role identities (adopted or ascribed) and their perceived compatibility will determine the success of the exchange and potential for perpetuating such relationships (Gursoy et al., 2002). Social exchange is found in the host-guest tourism exchange where stakeholders who benefit from tourism will support its development and seek to maximize its rewards (Sutton, 1967). The perceived structure of the destination tourism as a short-term visit and high-profit economic sector is strengthened by the perpetuation of a simulation of destination image – through media, movies, television screens and the internet – that fosters an illusion of predictable and recognizable completeness (Kroker et al., 1990). This view is disrupted when new forms of mobility, e.g., long-stay migrants come on the scene. New unexpected forms of 'difference' (Pfohl, 1994) raise the intensity of role identity expectations and cast doubt on the potential benefits of the social exchange.

The social and economic success of a tourism destination and its sustainability depends upon the directed identification, coordination, and organization of stakeholders (Gray, 1989; Reed, 1997). Stakeholders potentially include all members of a community regardless of differences in goals, interests or designs (Aas et al., 2005). They are specifically identified by their expressed or implied attitudes toward resource issues or 'the perceived personal benefit of tourism to an individual or dependence on it' (Andereck et al., 2005, p. 1061). Stakeholders are also identified in terms of how they are affected by tour operators' objectives or tourism operations (Ryan, 2002b). Whether from the observed personal subjectivity of the stakeholder or from their operational stance, the definition of a stakeholder is normative and specific to the destination (Freeman, 1984). In Jeju, the sentiment of residents toward Yemeni refugee migrants is a key stakeholder issue framed by the perceived benefits and outcomes of social exchange based on adopted or ascribed role identity.

### *Power and alienation*

Power relations shape the identity of stakeholders and can alter or preclude collaborative action (Reed, 1997), especially given that power is often asymmetrical (Hall, 1994). The issue of power is always at the forefront of discussions of economic or social consequences, and in tourism, it is observed in the relationship dynamics between stakeholders. Those involved in the junction between host and tourist may harbor ethnocentric feelings of distrust toward those who do not conform to existing forms of social exchange (Hollinshead, 1993). Those newcomers who fail to acculturate or to follow established normative expectations will lose cultural capital as their formulae of race, culture, economy, and language gradually come to be seen as 'undeveloped' or 'foreign' (Hunter, 2001). Anomic cultural relationships arise from a lack of shared social values and can result in a slippage of confidence in the value of social exchange, leading to entrenched power distance and alienation in a culturally diverse community.

Stakeholder identity and their relationships to power are largely determined by their degree of social and economic engagement. Stakeholder subjectivity is identity based and magnified (especially in smaller tourism destinations) by the proximity of government policy to real-world effects. These effects will include efforts to capitalize on the economic potential of tourism and the representation and commodification of destination culture for tourism promotion (Pritchard & Morgan, 2001; Schellhorn, 2010). Bad decision-making in destination tourism management deepen resentment among disenfranchised elements of the resident population and will divide the community. Good decision-making will seek consensus and the wider distribution of benefits derived from tourism, working to ensure the sustainability of cultural and natural resources. In a democracy, government policy is closely tied to public opinion. A public bias toward minority groups obstructs their political legitimacy and a sense of shared ownership in society (Bramwell & Sharman, 1999).

### *Trust and social responsibility*

Cultural sentiments of distrust in Jeju hinder progress in the migration-tourism nexus (Balli et al., 2016) and impede the benefits of migration that can work as a marketing tool, exposing and encouraging residents to visit migrants' home country or culturally related countries. Barriers to trust are rooted in the collective perception in which migrant communities are deviant in the sense that they fail to conform to dominant normative expectations of what it is to be a tourist. Refugee migrants in Jeju may fail to conform to cultural symbols and power issues that support represent normative social expectations (Hunter, 2001), relegating them – in the view of residents – to the ascribed identity of 'Other'. There may, however, be benefits to tourism destinations that host migrant populations. These potentially fringe stakeholders can contribute to a more inclusive attitude among residents toward their non-Korean community members (Khazaei et al., 2015). These migrants can also affect trade in goods and services between their home and host country, bringing to bear a beneficial economic influence on the destination (Tadesse & White, 2012).

Yemeni migrant arrivals may begin to build small firms or enterprises that will contribute to the community. Partnerships supported by tourism public policy can contribute to the competitiveness of the tourism sector (Thomas, 2007). At this point, however, the lack of trust in the public discourse towards these migrants eclipses expectations regarding inclusion in the short term. Understanding Korean residents' held subjectivities toward Yemeni migrants in Jeju is a first step toward building trust between cultures and exploiting in the most effective way their mutual roles in the continued development of the island's tourism economy (Khazaei et al., 2017). Trust, in the structure of new destination tourism forms, exists in the space between people in pursuit of beneficial social exchange, regardless of adopted or ascribed role identity (Simons, 1990). Trust is a performative event that takes place in intersubjective and collective experiences of nation-ness, community interest and the negotiation of cultural values (Bhabha, 1994). Trust requires a model of responsible tourism built on engagement that facilitates mutual learning between cultures (Young, 1996).

In this paper, the migration-tourism link is explored in the context of Yemeni refugee migrant arrival in Jeju. These refugee migrants, motivated by a search for asylum, face economic and cultural challenges in their destination community. Theoretical themes concerning social exchange and role identity, power and alienation, and trust and social responsibility have been employed to describe the social discourse surrounding Yemeni arrivals in Jeju. In the following section the operationalization of these theoretical themes using Q method is described.

## Methodology

This paper employs Q method as an inductive approach to identifying residents' operant subjectivities toward Yemeni refugee migrants in Jeju Island, South Korea based on the theoretical concourse (Hunter, 2014) described in the literature review. Operant subjectivity refers to functional rather than logical categorization (Brown, 1980) of attitudes and beliefs towards a public discourse (concourse) and subjective divisions within society (Bahl & Milne, 2006). Individuals might revise or change their opinion towards a discourse or the event it represents, but the clusters of subjectivity remain as a functional (Stephenson, 1978). Q method is designed to identify 'clusters of subjectivities' rather than the 'subject's subjectivity', employing both mathematical and interpretive tools (Stergiou & Airey, 2011) through a 5-step process:

1. Identify the theoretical concourse:
2. Generate a Q-set (a sample of 42 statements that represents the concourse):
3. Recruit a P-set (a sample of 30 respondents) to perform a Q-sort:
4. Perform Q factor analysis to identify clusters of subjectivity:
5. Perform interpretive analysis to theoretically position clusters.

In Q method, participants represent their subjective views by ranking a set of stimuli, for example, statements or photographs and after a factor analysis extracts common perspectives (Stephenson, 1993) which are then subject to interpretive analysis. Researchers in the fields of tourism and hospitality have utilized Q method to explore various topics in their studies. For instance, the method has been used to explore the meaning of the destination (Huang et al., 2017); performer's subjectivities in tourism culture parks (Hunter, 2014); influence of nationalism on tourist-host relationships (Griffiths & Sharpley, 2012); and residents' subjectivities toward tourism (Hunter, 2013).

### *Concourse development*

In Q method the concourse can be identified through interpretive content analysis (media text, interviews or any general representation of the public discourse) or it can be theoretically defined based on the review of academic literature (Stephenson, 1993). In this study, the latter approach has been adopted based on the migration-tourism link as defined by the three theoretical themes previously discussed. Q method is an inductive approach (McKeown & Thomas, 1988; Barry & Proops, 1999) and it should be noted that the basic contours of the concourse were indirectly noted by researchers in the early conceptual stages of the study starting in 2016 when Yemeni refugee migrants began to arrive in Jeju (No Cut News, 2018) until the writing of this paper began in 2019. The concourse and Q-set items were eventually designed through a double-blind coded process before determining the configuration of explanatory theoretical themes used in this study. The concourse is structured by three theoretical themes outlined as follows:

- Social exchange and role identity among stakeholders:
  - ^ Economy – how refugees offer an opportunity to generate wealth and social value in Jeju (Andereck et al., 2005) as members of the tourism economy (Narayan, 2004):
  - ^ Nationalism – how identity is based on allegiance to the State (Smith, 2009) frames subjective views on the migration-tourism link (Griffiths & Sharpley, 2012):
  - ^ Motivations – how the migration-tourism link works to influence destination choice (You, O'leary, Morrison, & Hong, 2000).

- Power and alienation:
  - ^ Xenophobia – how 'fear of the stranger', denoting a dislike of refugee newcomers (Harris, 2002) – especially those in an asymmetrical power position, receive hostility and resistance to cultural change (Ortona, 2017):
  - ^ Islamophobia – how this prejudice might block altruism in favor of short term benefits (Allen, 2007).

- Trust and social responsibility:
  - ^ Trust – how people treat each other will affect the degree of confidence in placing oneself in the hands of others (Pettit, 1995):
  - ^ Reciprocity – the expectation that people will actively respond to each other in reasonable and mutually dependent ways (Gouldner, 1960).

### *Q-set and P-set sampling and the Q-sort*

Q method identifies functional divisions within society surrounding a social discourse based on clusters of opinions, beliefs, and attitudes (Watts & Stenner, 2005). It employs two modes of sampling (Brown, 1980). First, the theoretical concourse identified in this study was sampled to produce a Q-set of 42 statements (Van Exel & Graaf, 2005) through a double-blind process of selection and elimination. The statements represent the three theoretical themes regarding the migration-tourism link discussed in the literature review. These statements were coded and printed on cards and presented to respondents for the Q-sort.

Second, respondents are recruited to form a P-set. A core group of individuals was found by their involvement in the refugee situation and additional respondents were found through a snowballing process. Thus purposive sampling approach was employed to ensure that respondents had an invested understanding of the issues represented in the concourse (Eden et al., 2005). Respondents were Jeju resident-stakeholders concerned with the issues surrounding the presence of Yemeni refugee migrants in Jeju and their perceived socio-economic impacts (Miles & Huberman, 1994). The P-set consisted of 30 respondents (Van Exel & Graaf, 2005; Hunter, 2014).

Previous Q studies have used P-sets of as many as 66 (Fairweather & Swaffield, 2001) and as few as 27 (Dewar et al., 2007) or 20 respondents (Hutson & Montgomery, 2006). In Q method, the Q-set should be larger than the P-set (Van Exel & Graaf, 2005). Respondents were contacted and invited to perform the Q-sort process, arranging the Q-set statements into piles from strongly disagree to strongly agree. The recruitment of the P-set and performance of the Q-sort were conducted in Korean.

### *Statistical and interpretive analysis*

The Q factor analysis was processed using dedicated Q method software to identify centroids or factors that identify clusters of subjectivity. Any means of performing a Q factor analysis can obtain these objectives, but in this study PCQ for Windows Standard Edition was employed. It is worth noting, before moving on to an in-depth discussion on the results that statistical tests for validity and reliability are not prevalent in Q. Rather, like other interpretive methods, the quality of findings is found through replication (Gobo, 2008). The factors were interpreted according to their relationship to the theoretically derived concourse and the results are discussed in the following section.

## Results

### *Descriptive statistics*

The P-set included 11 male and 19 female respondents who were Korean residents in Jeju. Age ranged from 17 to 67 years old with a mean of 42. Four factors were found in the 30 Q-sorts, after the Q factor analysis and graphic rotation process. They included: Factor A (accounted for 23% of the variance and included 12 sorts); Factor B (accounted for 19% of the variance and included 8 sorts); Factor C (accounted for 6% of the variance and included one sort); and Factor D (accounted for 5% of the variance and also included one sort). Nine sorts were confounded (significant in more than one factor) and one sort (7) was not significant. In the interpretive phase factors are identified as 'clusters of subjectivity'. In Q method, individuals might revise or change their opinion towards the concourse but the clusters of subjectivity (factors) within the discourse remain. Therefore, the number of respondents within a factor does not determine its value or significance but rather the number of clusters within the P-set is what is of interest (Hunter, 2014). The clusters and respondents' demographic information are summarized in Table 1.

### *Analysis and Interpretation*

The Q factor analysis identified four clusters of residents' subjectivity toward Yemeni refugee migrants in Jeju Island. Four consensus statements were found, items upon which all significant sorts agreed. Those items are depicted in Table 2, below. Respondents all mildly or strongly disagreed with Item 10, 'Koreans are not xenophobic toward foreigners'. This reflects all respondents' self-perception of Korea as an insular society unwilling to interact with a refugee migrant population. This suggests a disingenuous attitude toward power relations that precludes collaboration (Reed, 1997) and obstructs the political legitimacy of minority groups in Korean society

**Table 1.** Q sort factor analysis results.

22 Q sorts have been accounted for in 4 factors out of 5; 9 significant in two or more factors (confounded) and 1 factor not significant

| Sort | A (12) | B (8) | C (1) | D (1) | Descriptions* |
|---|---|---|---|---|---|
| 1 | 0.51 | | | | K M 50 Police officer; Protestant |
| 2 | 0.72 | | | | K F 48 Housewife; Not religious |
| 4 | 0.41 | | | | K M 49 Police officer; Protestant |
| 5 | 0.49 | | | | K F 56 Real estate agent; Protestant |
| 10 | 0.53 | | | | K F 28 Hotelier; Protestant |
| 12 | 0.68 | | | | K F 46 Professor; Protestant (inactive) |
| 14 | 0.52 | | | | K F 34 Ph.D. candidate, part-time employee; Protestant |
| 15 | 0.70 | | | | K F 25 Cook; Not religious |
| 16 | 0.47 | | | | K M 17 High School student; Not religious |
| 22 | 0.63 | | | | K F 31 Tour guide; Buddhist |
| 27 | 0.44 | | | | K M 24 University student; Not religious |
| 30 | 0.61 | | | | K F 48 Financial expert; Protestant (inactive) |
| 4 | | 0.40 | | | K M 49 Police officer; Not religious |
| 8 | | 0.61 | | | K F 47 Math tutor; Buddhist |
| 18 | | 0.67 | | | K F 42 Coffee shop owner; Catholic |
| 21 | | 0.71 | | | K M 61 Farmer; Protestant |
| 23 | | 0.50 | | | K F 23 University student; Protestant (inactive) |
| 24 | | 0.66 | | | K M 60 Taxi driver; Protestant |
| 25 | | 0.70 | | | K M 38 Restaurateur; Not religious |
| 26 | | 0.70 | | | K M 67 Retiree; Buddhist |
| 24 | | | 0.40 | | K M 60 Taxi driver; Protestant |
| 17 | | | | 0.42 | K M 35 Ph.D. candidate, researcher; Agnostic |
| Sorts 3, 6, 9, 11, 13, 19, 20, 28, and 29 confounded; Sort 7 not significant | | | | | |
| Confounded | | | | | |
| 3 | 0.52 | | 0.53 | | K F 47 Teacher; Presbyterian (inactive) |
| 6 | 0.41 | 0.46 | | | K F 49 Housewife; Protestant |
| 9 | 0.46 | | 0.46 | | K F 37 Tourism organization employee; Protestant |
| 11 | 0.42 | 0.75 | 0.53 | 0.40 | K F 34 Hotelier, university lecturer; Protestant (inactive) |
| 13 | 0.40 | 0.59 | | | K F 38 Company employee; Protestant |
| 19 | 0.40 | 0.46 | | 0.47 | K M 43 Police officer; Catholic |
| 20 | 0.40 | 0.56 | | | K M 25 University student, part-time employee; Not religious |
| 28 | 0.59 | | | | K F 52 Nurse; Catholic |
| 29 | 0.60 | | | | K F 39 Teacher; Not religious |

| Eigenvalue and % variance | | | | | | |
|---|---|---|---|---|---|---|
| 5 factors | 1 | 2 | 3 | 4 | 5 | Totals |
| Eigenvalue | 6.75 | 5.70 | 1.81 | 1.39 | 1.40 | 16.76 |
| % variance | 23 | 19 | 6 | 5 | 3 | 57 |

*Descriptions key: Nationality: K = Korean.
Gender: M = Male (N = 11), F = Female (N = 19).
Age: n = Age. (Range: 67 – 17; Mean: 42.6).
Short statement describing respondent (sorter) by occupation and religion.

**Table 2.** P set consensus and distinguishing statements with relative distributions.

| Item | P set consensus statements (4)<br>Distinguishing statement for cluster D (1) | Distribution score | | | |
|---|---|---|---|---|---|
| **Consensus statements, P set** | | A | B | C | D |
| 10 | Koreans are not xenophobic toward foreigners | −3 | −2 | −2 | −3 |
| 25 | It is easier to get a refugee status if you are not a Muslim | −1 | −1 | −1 | 0 |
| 33 | The influx of refugees from Yemen have a negative impact on tourism | 1 | 1 | 1 | 1 |
| 39 | There is a cultural exchange between local residents and international tourists | 0 | −1 | −1 | −1 |
| **Distinguishing statement for cluster D** | | A | B | C | D |
| 22 | Cheap airfare, as well as short travel time, prompted a rapid surge of Yemeni refugees | −1 | 0 | −1 | 4 |

(Bramwell & Sharman, 1999). The situation of Yemeni migrants in Jeju is one of the asymmetrical power relationships with society (Hall, 1994). The other three consensus statements, 25, 33 and 39 elicited mild or neutral responses. Among the four clusters, there was only one distinguishing statement, for Cluster D. This statement, 'cheap airfare, as well as short travel time, prompted a rapid surge of Yemeni refugees', elicited a 'strongly agree' (4) response from the respondent in contrast to the other clusters which seem to undervalue the structural cause of the appearance of Yemeni refugee migrants in Jeju.

In Table 3, the results of a deeper interpretive analysis (step 5 of the Q method process) is depicted. In this table, each of the four clusters is shown with a name. In Q, the interpretive process of naming the clusters is carried out by the researchers in order to provide the reader with an organic picture of their relative operant subjectivities. The naming is then clarified by a deeper interpretation of their representative statements. The clusters of subjectivity are named as follows: Cluster A – The Nation comes First; Cluster B – Yemenis are Wolves in Sheep's Clothing; Cluster C – Guarded Mistrust; and Cluster D – A Balanced Outlook. Each of these four clusters is described in terms of the ten statements most agreed and disagreed (scores of 4, −4, 3 and -3). These ten statements combined, express a cluster of subjectivity toward the concourse, in this case, the flow of communicability around the topic of Yemeni refugee migrants in Jeju and Koreans' operant subjectivity towards them. These statements are coded as follows:

- Social exchange and role identity among stakeholders (25 items):
  ^ Economy – items 29-42:
- ^ Nationalism – items 1-5:
- ^ Motivations – items 17-22:

- Power and alienation (7 items):
  ^ Xenophobia and Islamophobia – items 6-12:

- Trust and social responsibility (10 items):
  ^ Trust – items 23-28:
- ^ Reciprocity – items 13-16.

Cluster A (The Nation comes First) represents a concern for tourism and the economy (items 30 & 40) but at the expense of social exchange with newcomers, such as

**Table 3.** Characterizing statements for clusters using highest distribution scores.

| Item | Clusters with statements with highest distribution scores | Distribution Score |
|---|---|---|
| **Cluster A: The Nation comes First** | | |
| 30 | Tourism is important to the economy of Jeju Island | 4 |
| 40 | Jeju Island is attractive to tourists | 4 |
| 12 | Jeju Island has the highest foreign crime rate nationwide | −4 |
| 13 | I believe that more Yemenis should be granted refugee status in South Korea | −4 |
| 4 | The people of our nation should always come first | 3 |
| 20 | Yemenis chose Jeju Island over other destinations due to its visa-waiver program | 3 |
| 37 | Tourism improves the lives of local residents | 3 |
| 8 | Yemenis are terrorists in the guise of refugees | −3 |
| 10 | Koreans are not xenophobic toward foreigners | −3 |
| 29 | Since Korea lacks land, resources, and jobs we should not help others | −3 |
| **Cluster B: Yemenis are Wolves in Sheep's Clothing** | | |
| 23 | I do not trust Muslims | 4 |
| 5 | I am against having refugees in my country | 4 |
| 13 | I believe that more Yemenis should be granted refugee status in South Korea | −4 |
| 16 | We should show humanity to asylum seekers and return the kindness shown to us during the war | −4 |
| 8 | Yemenis are terrorists in the guise of refugees | 3 |
| 11 | I am worried about my own safety and my family's due to Yemenis | 3 |
| 19 | Yemenis are illegal immigrants who came to earn money | 3 |
| 27 | I believe that Yemenis will go back when the civil war is over | −3 |
| 28 | Some radicals are spreading Islamophobia through the false rumors | −3 |
| 42 | Refugees could help ease serious labor shortages in agriculture and fishery | −3 |
| **Cluster C: Guarded Mistrust** | | |
| 23 | I do not trust Muslims | 4 |
| 24 | I trust Christians, Catholics, and Buddhists more than Muslims | 4 |
| 27 | I believe that Yemenis will go back to when the civil war is over | −4 |
| 42 | Refugees could help ease serious labor shortages | −4 |
| 2 | We Koreans are one of the few racially pure nations | 3 |
| 34 | Local tourists are nicer and more considerate than foreign tourists | 3 |
| 35 | I have had bad experiences with international tourists | 3 |
| 13 | I believe that more Yemenis should be granted refugee status in South Korea | −3 |
| 15 | We should help Yemenis suffering from war; we also got help from foreign nations during the Korean War | −3 |
| 17 | Yemenis had to seek asylum in Jeju Island due to their Civil War | −3 |
| **Cluster D: A Balanced Outlook** | | |
| 20 | Yemenis chose Jeju Island over other destinations due to its visa-waiver program | 4 |
| 22 | Cheap airfare, as well as short travel time, prompted a rapid surge of Yemeni refugees | 4 |
| 1 | I am proud of being Korean | −4 |
| 27 | I believe that Yemenis will go back to when the civil war is over | −4 |
| 3 | I strongly believe that nationalism is wrong | 3 |
| 6 | The majority of Koreans are hostile toward Yemeni immigrants | 3 |
| 21 | Korea is an attractive destination due to safety, higher wages & more jobs available for foreign migrants | 3 |
| 10 | Koreans are not xenophobic toward foreigners | −3 |
| 11 | I am worried about my own safety and my family's due to Yemenis | −3 |
| 23 | I do not trust Muslims | −3 |

the Yemeni immigrant community (Cook et al., 2013). This view devalues opportunities to engage with this 'other' (items 12 & 13) and overlooks the benefits of reciprocity (Khazaei et al., 2017). There is a strong sense of national isolationism, suspicion or disengagement with 'the other' and a self-identity as being the sole stakeholder group in the Jeju economy (4, 8 & 10). Cluster A reflects a disinterest in social exchange with, or responsibility for, Yemeni refugee migrants.

Cluster B (Yemenis are Wolves in Sheep's Clothing) represents a powerful distrust toward the refugee migrants (items 23, 27 & 28) bordering on xenophobia or

Islamophobia (items 8 & 11) and nationalistic isolationism (item 5). A strong lack of sympathy or reciprocity is also indicated in these Q-sorts (items 13 & 16). This view reflects cultural sentiments that work in opposition to the potential benefits available from the migration-tourism link (Balli et al., 2016). In addition, the explicit bias against Yemenis in Jeju works in operant subjectivity to consciously and socially marginalize this group of refugees (Bramwell & Sharman, 1999).

Cluster C (Guarded Mistrust) shares Cluster B's powerful operant distrust of the Yemeni migrants (items 23, 24 & 27). But this cluster also reflects subjectivities toward migrants' roles in the tourism economy. It favors domestic visitors over international tourists – or migrants – (items 34 & 35) and sees not material, the social or psychological value in the labor potential of the new arrivals (item 42) (Andereck et al., 2005). This view rejects social engagement with Yemeni refugees (Gursoy et al., 2002) without being expressly Islamophobic (items 13, 15 & 17). This view simply seeks to avoid engagement with foreign visitors or new arrivals (Hunter, 2001).

Cluster D (Balanced Outlook) represents a minority view that is more concerned with the structural reality behind the arrival of Yemeni refugee migrants and how motivations rather than ideology shape their current condition (items 20, 21 & 22). This perspective favors focusing on the goals and designs of migrants rather than on nationalistic ideology as a barrier to social exchange (items 1 & 3).

## Discussion and conclusion

The purpose of this study was to employ Q method to understand residents' subjectivities toward Yemeni migrants to Jeju Island in South Korea. In the public discourse, there has been an ongoing discussion regarding the value or potential social problems associated with allowing this Muslim population into Korea for what appears to be an indefinite period of time. In previous research, in tourism geographies, a link between tourism and migration has been strongly established and in addition has been tied to the impacts of social movement on social capital (Casado-Díaz et al., 2014). In this study, four views on the refugee migrants in Jeju were found, and a summary of these views are shown in Table 4.

In the migration-tourism link, there is a distinction between voluntary and refugee migration. The key finding in this study is that all respondents identified with the viewpoint that Korea is an insular society unwilling to interact with a refugee migrant population. This unwillingness to engage 'the other' (the Yemeni migrant) raises concerns regarding the potential for social and economic success for these new arrivals and by extension, for the community (Reed, 1997), especially when social power between them is asymmetrical (Hall, 1994). Rather than vilifying Korean respondents for opposing refugees, it is important to recognize that such migrants, arriving in large groups, can pose a crisis rather than an opportunity since they typically lack the option to re-migrate (Williams et al., 2018). Residents lack any meaningful motivation to engage with this refugee migrant population.

The second key finding concerns a linkage between the theoretical themes of social exchange, power, and social responsibility. It is theorized that interactions between groups can generate mutual benefits (Gursoy et al., 2002). However, findings suggest

**Table 4.** Operant subjectivities by cluster.

| Operant subjectivities by cluster expressed using declarative statements |
|---|
| **Cluster A: The Nation comes First** |
| Jeju Island is an attractive tourism destination and tourism is important to the island economy. It improves the lives of residents; Jeju is safe and crime free. Yemen refugees have chosen Jeju due to its visa-waiver program. Although they are not terrorists, any more immigrants should not be granted refugee status. Korea does not lack land, resources and jobs but the interests of the nation should always come first. |
| **Cluster B: Yemenis are Wolves in Sheep's Clothing** |
| Muslims cannot be trusted and no more Yemenis should be granted refugee status in our country. They are terrorists in the guise of refugees and are dangerous. Yemenis are here for money and they are a burden to the economy. Once here they will never return to their own country. Korea has no obligation to show humanity to asylum seekers based on past events. |
| **Cluster C: Guarded Mistrust** |
| Muslims cannot be trusted; Yemenis are not here to escape their own civil war. They are a burden to the economy and will never leave. We do not owe them anything because of past historical events, like the Korean War. Korea is racially pure and our own people make the best tourists. We frequently have problems with international tourists in general. |
| **Cluster D: A Balanced Outlook** |
| I am not proud to be Korean because I believe nationalism is wrong and cultivates xenophobia. I do not distrust Muslim immigrants and am not afraid of them. But Yemenis in Jeju will never go back to their own country. They are here because Korea is an attractive and safe destination that offers job opportunities and higher wages for migrants. The Yemeni immigrants are here because cheap airfare was made available and Korea itself has offered a visa-waiver program to them. |

that although there are concerns for the economy, the potential for migrant contributions are overlooked (Cluster A – The Nation comes First), rejected outright (Cluster B – Yemenis are Wolves in Sheep's Clothing) or simply dismissed (Cluster C – Guarded Mistrust). Residents in Jeju may harbor the opinion that new refugee migrants are unable to acculturate or conform to existing forms of social exchange (Hollinshead, 1993) because they are simply too foreign (Hunter, 2001). This pragmatic, rather than isolationist outlook explains why residents seem to reject social responsibility for Yemeni migrants and overlook their potential to contribute to the destination economy (Tadesse & White, 2012). This may be exacerbated by a sense of the asymmetric social power between residents and migrants (Bramwell & Sharman, 1999). However, the theoretical theme of trust is also at work in the resident-refugee relationship. Cluster B – Yemenis are Wolves in Sheep's Clothing, one of two majority viewpoints, represents a strong distrust toward the Yemenis and borders on being Islamophobic, threatening progress in the acculturation or exploitation of the benefit potential of multiculturalism in Jeju (Khazaei et al., 2015).

Only Cluster D – A Balanced Outlook, a minority viewpoint recognized the structural realities surrounding the conditions that brought Yemeni migrants to Jeju. These included a visa-waiver program, short travel time and cheap airfare. If Koreans are strongly opposed to this community they can change existing policy to prevent more Yemenis from coming to Jeju. However, as expressed in this viewpoint, those Yemenis who are already here may be unlikely to go back when or if their civil war is over. So for those who are already here, Koreans might learn to 'trust Muslims', overcome Islamophobia and find ways to engage this new community as fringe stakeholders.

The dominant anti-refugee sentiments found among Jeju residents in this study threaten any potential for a sustainable resident-migrant relationship. Social sustainability relies on community prosperity and wellbeing and requires local strategic management and value creation (Ioannides, 1995; Ryan, 2002a). In this context, it also relies on residents' goodwill towards migrants (Yoon et al., 2001). However, refugee migrants arrive in large homogeneous groups without any apparent social or economic capital and without a cultural understanding of the destination. They also have little if any option for re-migration (Ivanov & Stavrinoudis, 2018). They offer little if any real value to the destination community (Andereck et al., 2005). In order to secure any social sustainability in the resident-refugee relationship in Jeju, government or private intervention is urgently needed.

### *Practical implications*

Findings indicate the dominant held beliefs among Jeju residents are associated with an unwillingness to interact with Yemeni refugee migrants. Residents undervalue migrants' social and economic potential. A dominant belief is that migrants pose a threat to the community and cannot be trusted. In terms of practical implications, it is important to note that these four clusters of operant subjectivity indicate points of view rather than individual opinions and that personal views can change over time. Operant subjectivity can change over time by implementing structural or attitude-related actions as follows:

- Structural changes – if more Yemeni or other migrant groups are unwanted, visa-free policies and cheap airfare promotion should be more carefully implemented. The promotion of destination tourism for economic growth should be tempered with a mature knowledge of all potential social ramifications.
- Attitude changes – for those who are already here, Jeju residents may need to change their subjective outlook from one of exclusion and distrust or fear to one of social exchange and engagement. New migrant populations should not be isolated but rather included as 'fringe stakeholders' in order to capitalize on their social, cultural and economic potential. Acculturation skills and cultural awareness education can be implemented through the government of privately funded workshop programs for Yemeni refugee migrants and for resident stakeholders.

### *Theoretical implications*

Previous research has demonstrated how voluntary migration can contribute to destination economies in the form of residential investment, exchange education and voluntary labor (Montezuma & McGarrigle, 2019). The key to its success is that this type of migrant has relative freedom to re-migrate. Refugee migration has also been examined but largely from a descriptive point of view in which the problems associated with the crisis are described and the potential benefits are contested (Dibeh et al., 2019). The empirical benefits of refugee migration to the destination are hypothetical, in terms of tapping fringe stakeholder potential (Schellhorn, 2010) or their ability to bear a positive economic influence on the destination by tapping their home country resources through trade (Tadesse & White, 2012).

This study has produced equally pessimistic findings by identifying a widely held public ambivalence to the refugee immigrant. However, this study offers a contribution to tourism research on the migration-tourism link by positioning it within a set of theoretical themes. It was found that social exchange, power and alienation, and social responsibility are valuable theoretical paths toward finding empirical solutions to fully exploiting the potential of refugee migrants for the benefit of the destination community.

### *Limitations and future research*

The limitation of this study is that it has outlined the contours of the refugee migrant condition in tourism research without finding the potential benefits of the phenomenon in Jeju Island. Residents fail to see any social or economic value in the Yemeni migrants. In future research, the viewpoints or actual social condition of refugee migrants must be explored in terms of their social values, goals and objectives and skill sets. They seem to be largely viewed as objects, or as a 'problem' rather than as individuals with a voice. Future research should give refugee migrants their voice by exploring their needs and motivations. In addition, the viewpoints of voluntary migrants and other non-residents toward refugee migrants at a destination are largely unexplored.

Another limitation unique to Q is in its exploratory nature. The sampling involved in the derivation of Q-set items from the concourse does not necessarily result in a symmetrical representation of the theoretical themes in the Q-set. In addition, the purposive sampling for a P-set of respondents will not proportionately represent the found clusters of subjectivity. In this sense, the match between Q-set and P-set is a delicate balance. Q-method driven studies will always benefit from replication and fine tuning in the same or comparative research contexts.

## Disclosure statement

No potential conflict of interest was reported by the author(s).

## References

Aas, C., Ladkin, A., & Fletcher, J. (2005). Stakeholder collaboration and heritage management. *Annals of Tourism Research*, *32*(1), 28–48. doi:10.1016/j.annals.2004.04.005

Allen, C. (2007). Islamophobia and its consequences. In *European Islam. Challenges for Public Policy and Society*, 144–167. CEPS Centre for European Policy Studies, Paperbacks, Brussels.

Andereck, K. L., Valentine, K. M., Knopf, R. C., & Vogt, C. A. (2005). Residents' perceptions of community tourism impacts. *Annals of Tourism Research, 32*(4), 1056–1076.

Bahl, S., & Milne, G. R. (2006). Mixed methods in interpretive research: an application to the study of the self concept. In R. W. Belk, (Ed.) *Handbook of qualitative research methods in marketing*, 198–218. Edward Elgar Publishing Limited.

Balli, F., Balli, H. O., & Louis, R. J. (2016). The impacts of immigrants and institutions on bilateral tourism flows. *Tourism Management, 52*, 221–229. doi:10.1016/j.tourman.2015.06.021

Barry, J., & Proops, J. (1999). Seeking sustainability discourses with Q methodology. *Ecological Economics, 28*(3), 337–345. doi:10.1016/S0921-8009(98)00053-6

Bhabha, H. K. (1994). *The location of culture*. Routledge. doi:10.4324/9780203820551

Boyle, P. J., Halfacree, K. H., & Robinson, V. (2014). *Exploring contemporary migration*. Routledge.

Bramwell, B., & Sharman, A. (1999). Collaboration in local tourism policymaking. *Annals of Tourism Research, 26*(2), 392–415. doi:10.1016/S0160-7383(98)00105-4

Brown, S. R. (1980). *Political subjectivity: Applications of Q methodology in political science*. Yale University Press.

Casado-Díaz, M. A., Casado-Díaz, A. B., & Casado-Díaz, J. M. (2014). Linking tourism, retirement migration and social capital. *Tourism Geographies, 16*(1), 124–140. doi:10.1080/14616688.2013.851266

Cook, K. S., Cheshire, C., Rice, E. R., & Nakagawa, S. (2013). Social exchange theory. In *Handbook of social psychology*, 61–88. Springer.

Dewar, K., Li, W. D., & Davis, C. H. (2007). Photographic images, culture, and perception in tourism advertising. *Journal of Travel & Tourism Marketing, 22*(2), 35–44.

Dibeh, G., Fakih, A., & Marrouch, W. (2019). Tourism–growth nexus under duress: Lebanon during the Syrian crisis. *Tourism Economics, 26*(3), 353–370. doi:1354816619836338

Eden, S., Donaldson, A., & Walker, G. (2005). Structuring subjectivities? Using Q methodology in human geography. *Area, 37*(4), 413–422.

El Aouni, F., Cascón-Pereira, R., & Hernández-Lara, A. B. (2013). The role of emigrants in the construction of a destination brand: a new research line. *Tourism and Hospitality Management, 19*(1), 35–47.

Eum, I, (2017). Korea's response to Islam and Islamophobia: Focusing on veiled Muslim women's experiences. *Korea Observer - Institute of Korean Studies, 48*(4), 825–849. doi:10.29152/KOIKS.2017.48.4.825

Fairweather, J. R., & Swaffield, S. R. (2001). Visitor experiences of Kaikoura, New Zealand: An interpretive study using photographs of landscapes and Q method. *Tourism Management, 22*(3), 219–228.

Filimonau, V., & Mika, M. (2019). Return labour migration: an exploratory study of Polish migrant workers from the UK hospitality industry. *Current Issues in Tourism, 22*(3), 357–378. doi:10.1080/13683500.2017.1280778

Freeman, R. E. (1984). *Strategic management: A stakeholder approach*. Pitman.

Gardner, R., Karakaşoğlus, Y., & Luchtenberg, S. (2008). Islamophobia in the media: A response from multicultural education. *Intercultural Education, 19*(2), 119–136. doi:10.1080/14675980801889658

Gobo, G. (2008). Re-conceptualizing generalization: Old issues in a new frame. In P. Alasuutari, L. Bickman, & J. Brannen (Eds.), *The Sage handbook of social research methods.*, 193–213. SAGE. doi:10.4135/9781446212165.n12

Gouldner, A. W. (1960). The norm of reciprocity: A preliminary statement. *American Sociological Review, 25*(2), 161–178. doi:10.2307/2092623

Gray, B. (1989). *Collaboration finding common ground for multi-party problems*. Josey Bass.

Griffiths, I., & Sharpley, R. (2012). Influences of nationalism on tourist-host relationships. *Annals of Tourism Research, 39*(4), 2051–2072. doi:10.1016/j.annals.2012.07.002

Gursoy, D., Chi, C. G., & Dyer, P. (2010). Locals' attitudes toward mass and alternative tourism: The case of Sunshine Coast. *Journal of Travel Research, 49*(3), 381–394. doi:10.1177/0047287509346853

Gursoy, D., Jurowski, C., & Uysal, M. (2002). Resident attitudes: A structural modeling approach. *Annals of Tourism Research, 29*(1), 79–105. doi:10.1016/S0160-7383(01)00028-7

Hall, C. M. (1994). *Tourism and politics: Policy, power and place*. Chichester.

Harris, B. (2002). Xenophobia: A new pathology for a new South Africa. In Hook, D and Eagle, G. (Eds.). Psychopathology and Social Prejudice. 169–184, *University of Cape Town Press, Cape Town*.

Hollinshead, K. (1993). Ethnocentrism in Tourism. In M. Khan, M. Olsen, & T. Var (Eds.). *VNR's Encyclopedia of hospitality and tourism*, 652–662, Van Nostrand Reinhold.

Hollinshead, K. (2007). Indigenous Australia in the bittersweet world: The power of tourism in the projection of 'old' and 'fresh' visions of aboriginality. In D. Butler & T. Hinch (Eds.), *Tourism and indigenous peoples: Issues and implications*., 281–304. Elsevier.

Huang, Y., Qu, H., & Montgomery, D. (2017). The meanings of destination: A Q method approach. *Journal of Travel Research*, *56*(6), 793–807. doi:10.1177/0047287516663652

Hudson, S., So, K. K. F., Li, J., Meng, F., & Cárdenas, D. (2019). Persuading tourists to stay–Forever! A destination marketing perspective. *Journal of Destination Marketing & Management*, *12*, 105–113. doi:10.1016/j.jdmm.2019.02.007

Hunter, W. C. (2001). Trust between culture: The tourist. *Current Issues in Tourism*, *4*(1), 42–67. doi:10.1080/13683500108667881

Hunter, W. C. (2013). Understanding resident subjectivities toward tourism using Q method: Orchid Island. *Journal of Sustainable Tourism*, *21*(2), 331–354. doi:10.1080/09669582.2012.685175

Hunter, W. C. (2014). Performing culture at indigenous culture parks in Taiwan: Using Q method to identify the performers' subjectivities. *Tourism Management*, *42*, 294–304. doi:10.1016/j.tourman.2013.12.014

Hutson, G., & Montgomery, D. (2006). How do outdoor leaders feel connected to nature places? A Q method inquiry. *Journal of Outdoor and Environmental Education*, *10*(2), 29–39. doi:10.1007/BF03400837

Illés, S., & Michalkó, G. (2008). Relationships between international tourism and migration in Hungary: Tourism flows and foreign property ownership. *Tourism Geographies*, *10*(1), 98–118. doi:10.1080/14616680701825271

Ioannides, D. (1995). A flawed implementation of sustainable tourism: The experience of Akamas. *Tourism Management*, *16*(8), 583–592. doi:10.1016/0261-5177(95)00081-X

Ivanov, S., & Stavrinoudis, T. A. (2018). Impacts of the refugee crisis on the hotel industry: Evidence from four Greek islands. *Tourism Management*, *67*, 214–223. doi:10.1016/j.tourman.2018.02.004

Jamass, M. M. (2014). *Images and perceptions of Muslims and Arabs in Korean popular culture and society* [Master's Thesis]. Miami: Florida International University.

Janta, H., Brown, L., Lugosi, P., & Ladkin, A. (2011). Migrant relationships and tourism employment. *Annals of Tourism Research*, *38*(4), 1322–1343. doi:10.1016/j.annals.2011.03.004

Jeon, D. (2011). The consumption of the diabolic Arab image in "oriental" Korea: "The other" enjoying the image of "the other. *Situations*, *5*, 99–109.

Khazaei, A., Elliot, S., & Joppe, M. (2015). An application of stakeholder theory to advance community participation in tourism planning: The case for engaging immigrants as fringe stakeholders. *Journal of Sustainable Tourism*, *23*(7), 1049–1062. doi:10.1080/09669582.2015.1042481

Khazaei, A., Elliot, S., & Joppe, M. (2017). Fringe stakeholder engagement in protected area tourism planning: Inviting immigrants to the sustainability conversation. *Journal of Sustainable Tourism*, *25*(12), 1877–1894. doi:10.1080/09669582.2017.1314485

Kroker, A., Kroker, M., & Cook, D. (1990). PANIC USA: Hypermodernism as America's. *Social Problems*, *37*(4), 443–459. doi:10.1525/sp.1990.37.4.03a00030

Kwon, J. (2019). South Korea's "Yemeni Refugee Problem." Middle East Institute. Retrieved February 14, 2020, from https://www.mei.edu/publications/south-koreas-yemeni-refugee-problem.

Lusticky, M., & Bednarova, M. (2018). Tourism destination competitiveness assessment: Research & planning practice. *Global Business and Finance Review*, *22*, 1–7.

McKeown, B., & Thomas, D. (1988). *Q methodology*. Sage Publications.

Miles, M. B., & Huberman, A. M. (1994). *Qualitative data analysis*. Sage.

Montezuma, J., & McGarrigle, J. (2019). What motivates international homebuyers? Investor to lifestyle 'migrants' in a tourist city. *Tourism Geographies*, *21*(2), 214–234. doi:10.1080/14616688.2018.1470196

Moufakkir, O. (2014). What's immigration got to do with it? Immigrant animosity and its effects on tourism. *Annals of Tourism Research*, *49*, 108–121. doi:10.1016/j.annals.2014.08.008

Nagel, C. (2016). Southern hospitality? Islamophobia and the politicization of refugees in South Carolina during the 2016 election season. *Southeastern Geographer*, *56*(3), 283–290. doi:10.1353/sgo.2016.0033

Narayan, P. K. (2004). Economic impact of tourism on Fiji's economy: Empirical evidence from the computable general equilibrium model. *Tourism Economics*, *10*(4), 419–433. doi:10.5367/0000000042430971

Ndione, L., Decrop, A., & Rémy, E. (2018). Migrants going back homeland for holidays: Rituals and practices of Senegalese migrants in France. *Annals of Tourism Research*, *70*, 25–38. doi:10.1016/j.annals.2018.02.004

No Cut News (2018). Jeju Yemeni refugee crisis, what have we learned? (In Korean) Retrieved January 3, 2020, from https://www.nocutnews.co.kr/news/5082072.

Ortona, G. (2017). Xenophobia is really that: A (rational) fear of the stranger. *Mind & Society*, *16*(1-2), 37–49. doi:10.1007/s11299-016-0190-7

Pêcheux, M. (1982). *Language, Semantics and Ideology: Stating the Obvious*. (H. Nagpal, Trans.), Macmillian.

Pettit, P. (1995). The cunning of trust. *Philosophy<Html_Ent Glyph="@Amp;" Ascii="&"/> Public Affairs*, *24*(3), 202–225. doi:10.1111/j.1088-4963.1995.tb00029.x

Pfohl, S. (1994). *Images of Deviance and Social Control: A Sociological History*. McGraw-Hill.

Prayag, G., Hosany, S., Nunkoo, R., & Alders, T. (2013). London residents' support for the 2012 Olympic Games: The mediating effect of overall attitude. *Tourism Management*, *36*, 629–640. doi:10.1016/j.tourman.2012.08.003

Pritchard, A., & Morgan, N. J. (2001). Culture, identity and tourism representation: Marketing Cymru or Wales? *Tourism Management*, *22*(2), 167–179.

Reed, M. G. (1997). Power relations and community-based planning. *Annals of Tourism Research*, *24*(3), 566–591. doi:10.1016/S0160-7383(97)00023-6

Ryan, C. (2002a). Equity, management, power sharing and sustainability: Issues of the "New Tourism. *Tourism Management*, *23*(1), 17–26. doi:10.1016/S0261-5177(01)00064-4

Ryan, C. (2002b). Tourism and cultural proximity: Examples from New Zealand. *Annals of Tourism Research*, *29*(4), 952–971. doi:10.1016/S0160-7383(02)00006-3

Schellhorn, M. (2010). Development for whom? Social justice and the business of ecotourism. *Journal of Sustainable Tourism*, *18*(1), 115–135. doi:10.1080/09669580903367229

Sheikh, F. (2019a). Exploring "Korean Islam" in a climate of exclusion and Islamophobia. *International Journal of Diaspora &Cultural Criticism*, *9*(2), 196–215. doi:10.15519/dcc.2019.07.9.2.196

Sheikh, F. (2019b). Exploring how mobility affects Muslim lives: The case of Yemeni refugees on Jeju Island. *International Journal of Diaspora&Cultural Criticism*, *9*(1), 70–99. doi:10.15519/dcc.2019.01.9.1.70

Simons, H. W. (Ed.). (1990). *The rhetorical turn: Invention and persuasion in the conduct of inquiry*. University of Chicago Press.

Smith, A. D. (2009). *Ethno-symbolism and Nationalism: A cultural approach*. Routledge.

Stephenson, M. L., & Ali, N. (2010). Chapter 15: Tourism and Islamophobia. In *Tourism in the Muslim World*. (pp. 235–251). Emerald Group Publishing Limited.

Stephenson, W. (1978). Concourse theory of communication. *Communication*, *3*, 21–40.

Stephenson, W. (1993). Introduction to Q methodology. *Operant Subjectivity*, *17*, 1–13.

Stergiou, D., & Airey, D. (2011). Q methodology and tourism research. *Current Issues in Tourism*, *14*(4), 311–322. doi:10.1080/13683500.2010.537316

Sutton, W. A. (1967). Travel and understanding: Notes on the social structure of touring. *International Journal of Comparative Sociology*, *8*(2), 218–223. doi:10.1177/002071526700800206

Tadesse, B., & White, R. (2012). Do immigrants enhance international trade in services? The case of US tourism services exports. *International Journal of Tourism Research, 14*(6), 567–585. doi: 10.1002/jtr.857

Thomas, R. (2007). Tourism partnerships and small firms: Power, participation and partition. *The International Journal of Entrepreneurship and Innovation, 8*(1), 37–43. doi:10.5367/000000007780007344

Tsartas, P., Kyriakaki, A., Stavrinoudis, T., Despotaki, G., Doumi, M., Sarantakou, E., & Tsilimpokos, K. (2019). Refugees and tourism: A case study from the islands of Chios and Lesvos, Greece. *Current Issues in Tourism*, Published Online, 1–17. doi:10.1080/13683500.2019.1632275

UNWTO (2010). *Tourism and Migration – Exploring the Relationship between Two Global Phenomena.* Retrieved August, 17–2019, from https://www.e-unwto.org/doi/book/10.18111/9789284413140

Van Exel, J., D., Graaf, G. (2005). Q methodology: a sneak preview. Retrieved March 27, 2019, from http://www.qmethodology.net/index.php?page=1&year=2005.

Watts, S., & Stenner, P. (2005). Doing Q methodology: Theory, method and interpretation. *Qualitative Research in Psychology, 2*(1), 67–91. doi:10.1191/1478088705qp022oa

Williams, A. M., & Hall, C. M. (2000a). Guest editorial: Tourism and migration. *Tourism Geographies, 2*(1), 2–4. doi:10.1080/146166800363411

Williams, A. M., & Hall, C. M. (2000b). Tourism and migration: New relationships between production and consumption. *Tourism Geographies, 2*(1), 5–27. doi:10.1080/146166800363420

Williams, A. M., Jephcote, C., Janta, H., & Li, G. (2018). The migration intentions of young adults in Europe: A comparative, multilevel analysis. *Population, Space and Place, 24*(1), e2123. doi: 10.1002/psp.2123

Williams, A. M., King, R., Warnes, A., & Patterson, G. (2000). Tourism and international retirement migration: New forms of an old relationship in southern Europe. *Tourism Geographies, 2*(1), 28–49.

Worster, W. T. (2012). The evolving definition of the refugee in contemporary international law. *Berkeley J. Int'l L, 30*, 94.

Yoon, Y., Gursoy, D., & Chen, J. S. (2001). Validating a tourism development theory with structural equation modelling. *Tourism Management, 22*(4), 363–372.

You, X., O'leary, J., Morrison, A., & Hong, G.-S. (2000). A cross-cultural comparison of travel push and pull factors: United Kingdom vs. *Japan. International Journal of Hospitality & Tourism Administration, 1*(2), 1–26. doi:10.1300/J149v01n02_01

Young, R. (1996). *Intercultural Communication: Pragmatics, genealogy, deconstruction.* Philadelphia.

# "Traditional Mexican Midwifery" tourism excludes indigenous "others" and threatens sustainability

Rosalynn A. Vega

**ABSTRACT**

Drawn by the allure of "ancient cultures," tourists inadvertently consume deauthenticated indigenous practices, including ethnomedical traditions such as midwifery. This is especially true in the case of "Traditional Mexican Midwifery" since stark differences exist between how midwifery practices unfold in indigenous contexts and how they are represented to global tourists. "Traditional Mexican Midwifery" tourism is a unique lens for examining some of the underlying, intersectional issues threatening "sustainability" in ethnomedical tourism. When nonindigenous individuals position themselves as representatives of "Traditional Mexican Midwifery" and indigenous midwives are excluded from profit chains, this type of tourism not only fails to meet the needs of the indigenous people it fetishizes, it actually excludes them from the equation entirely. While overt prejudice and racial discrimination of the colonial era no longer exist, neocolonial tourism in the form of "Traditional Mexican Midwifery" tourism subtly reinforces romanticized stereotypes of the indigenous "Other." Ethnographic examples, gleaned from multi-sited research in Mexico and Brazil, demonstrate the usefulness of "Traditional Mexican Midwifery" for critiquing existing misapplications of responsible tourism and proposing more sustainable futures. Tourism can be an indigenous process (in which local residents both motivate and reap the benefits of tourism), or it can serve a neocolonial agenda (in which tourism development and profits unequitably favor transnational elites). Current forms of "Traditional Mexican Midwifery" tourism are unsustainable due to intersectional exclusion of indigenous women based on race, class, and gender. To dismantle intersectional inequality in "Traditional Mexican Midwifery," coordinated efforts by Mexican Secretary of Health's Traditional Medicine and Intercultural Development (TMID), Mexican Midwifery Association, and critical tourism scholars and activists are required.

摘要

游客们被"古老文化"的魅力所吸引，不经意间消费了一些不被认可的本土习俗，包括诸如接生婆之类的民族医学传统。"传统的墨西哥接生婆"的案例中尤其如此，因为接生婆的做法如何在本土环境中展现，以及如何呈现给全球游客之间存在明显的差异。"传统的墨西哥接生婆"旅游是一个独特的视角，以检查一些潜在的，跨部门的问题，而这些问题威胁着民族医学旅游的可持续发展。当非原住民将自己定位为"传统墨西哥接生婆"的代表，而真正的原住民接生婆被排除在利润链之外时，这种类型的旅游不仅不能满足于它所迷恋的原住民的需求，实际上还将他们完全排除在外。虽然殖民时代的公开偏见和种族歧视已不复存在，但新殖民主义旅游以"传统墨西哥接生婆"的形式，微妙地强化了对原住民"他者"的浪漫化刻板印象。从墨西哥和巴西多个地点的研究中收集的民族志实例，表明"传统的墨西哥接生婆"在批评现有对负责任旅游不当应用和提出更可持续的未来方面的作用。旅游业可以是一个本土化的过程(在这个过程中，当地居民既能激发旅游的积极性，又能获得旅游的好处)，也可以充当新殖民主义的马前卒(在这个过程中，旅游业的发展和利润不公平地偏袒跨国精英)。由于基于种族、阶级和性别的跨部门排斥土著妇女，当前的"传统墨西哥接生婆"旅游形式是不可持续的。为了消除"传统墨西哥接生婆"中的部门间不平等，需要墨西哥卫生部、传统医学和跨文化发展部(TMID)、墨西哥接生婆协会和批判旅游学者和活动人士的共同努力。

## Introduction

While traditional Mexican midwifery is deeply rooted in Mexico's indigenous cultures, these practices have recently become a draw for tourists globally. The version tourists consume is often deauthenticated, as individuals offering themselves as representatives of these practices do not belong to the communities from which the practices originate. I have placed quotation marks around "Traditional Mexican Midwifery" when referring to this type of tourism to highlight the stark differences between how midwifery practices unfold in indigenous contexts and how they are represented to global tourists. In essence, traditional practices have been usurped from their original contexts and reconfigured for tourists' consumption.

This article provides ethnographic examples of how "Traditional Mexican Midwifery" tourism is unfolding across Mexico, and how it has been "offshored" to Brazil, tracing the multiple ways Mexican midwives and indigenous women are excluded from such tourism and its profit chains. These unequal power structures evidence colonial legacies continuing to unfold in Mexico today, suggesting "Traditional Mexican Midwifery" may be neocolonial in nature.

By questioning whether "Traditional Mexican Midwifery" tourism is an example of neocolonial tourism, I am not only querying the economic impacts (or lack thereof) on the "destination region," but also its potential effects on "race" and class inequalities. Since tourism is a *process*, this article examines the directional flows of ideas, people, and capital. Has "Traditional Mexican Midwifery" tourism been developed to meet the needs of the Mexican midwives it represents? I argue it not only fails to meet the needs of the indigenous people it fetishizes (from an extended distance), it actually excludes them from the equation entirely.

This article is about multiple, contingent *intersectional racialization processes* (Crenshaw, Gotanda, Peller, & Thomas, 1995; Crenshaw, 2014; Bridges, 2011; De Genova, 2005). I extend Kimberlé Crenshaw's concept of "intersectionality" to explore ways in which race makes class and class makes race. Intersectionality is a conceptual tool for understanding how different dimensions of inequality co-construct one another (Granzka, 2014). I use intersectionality as a way of thinking about the problem of sameness and difference in relation to power, a foundational logic of interlocking oppressions, and an examination of systemic domination overlapping race, gender, and socioeconomic class (see Yuval Davis, 2006; Granzka, 2014; Cohen, 1997). Merging Crenshaw's concept of "intersectionality" with Bridges' perspective on pregnancy as a "site of racialization" and "race as a process," I draw attention to the ongoing, interfacing processes leading to the material (re)realization of racial inequality.

Elsewhere, I use an intersectional approach to problematize notions of "empowerment" and "free choice" (Vega, 2018). Interest among nonindigenous individuals in "Traditional Mexican Midwifery" is intimately intertwined with the humanized birth movement in Mexico—a movement inspired by alternative birth in the Global North (see below). Empowerment is so central to the ideological underpinning of the humanized birth movement that the 2014 World Week for Humanized Birth's slogan is, *"Parir es poder: empoderate"* ("To Give Birth is to Be Empowered: Empower Yourself," my translation). My intersectional approach reveals, contrary to the intentions of multiple public health programs in Mexico, hegemony is reinscribed when transnational midwives "empower" indigenous midwives living in rural settings with ideas from the Global North. Labeling some women as "empowered" and others "disempowered" produces a false binary, obscuring more about agency and dynamism than it reveals. Instead, an intersectional approach allows for nuances to emerge from multiple intersections among gender, race, and class.

An advantage of intersectional approaches is insights gleaned are place-based and specific to the study population. With these insights, I argue the notion of "empowerment" is based on neoliberal values originating in the Global North (e.g., individualism and choice). For most Mexican women, choices are either nonexistent or prestructured. The very idea of "choice" emerges from a middle- and upper-class perspective—"choice" is a privilege of those affording private medical services.

This article builds upon prior publications. I use "Traditional Mexican Midwifery" tourism as a unique lens for examining underlying, intersectional issues threatening "sustainability" in ethnomedical tourism. Analyzing the physical and social mobility of some individuals and the relative immobility of others, I argue "Traditional Mexican Midwifery" is unsustainable due to the intersectional exclusion of indigenous women based on race, class, and gender.[1]

This study uses the lens of "Traditional Mexican Midwifery" tourism to offer an overarching critique regarding responsible tourism. This critique can be applied to a range of medical tourism practices, along with "ethnic" tourism practices more generally. I propose reconfiguring responsible tourism to emphasize the cultural aspects of sustainability. Responsible tourism, when applied to medical tourism, holds potential for (re)defining social sustainability.

## Humanized birth vs. traditional birth

Focus is primarily on two contrasting types of birth: "humanized birth" (a type of alternative birth that has become a marker of cultural capital among affluent Mexican couples) and "traditional birth" (a type of alternative birth practiced among indigenous communities for generations). Both humanized and traditional births represent alternatives to medicalized births the majority of Mexican women experience when birthing in a hospital setting. Couples having a humanized birth value their ability to choose the particular professional attending the birth, which family members will be present, which physical positions they prefer for various stages of the birthing process, and the type of environment in which they will give birth. In contrast, as recipients of conditional cash transfers, women who have traditional births are actively defying Mexican Secretary of Health's mandate that *all* births take place in hospitals since *every* birth can potentially involve life-threatening complications (Vega, 2017b, 2018). Their births are less about choosing details of the birth and more about resistance to the subjugation of their bodies (Foucault, 1990[1978]).

By studying humanized and traditional births together, I bring unequal power structures into stark relief. Nader (1972) critiques the tendency among anthropologists to denounce unequal societal power structures through ethnographic observations of marginalized, disempowered, and lower-class individuals. This tendency is considered "studying down" because anthropologists are casting their analytical gaze down toward the bottom of the socioeconomic spectrum. She suggests anthropologists harness the ethnographic method to shift their gaze toward the top of the socioeconomic spectrum, thus opening up new possibilities for critically analyzing how inequality is produced from within seats of power. By "studying up"—conducting ethnographic research within spheres of affluence—I am heeding Nader's 1972 argument. The alternative births I describe are accessible only to the wealthiest Mexican couples. The costs associated with a midwife-assisted alternative birth, including prenatal care, homeopathic treatments, and biodynamic sessions, are 31,600 pesos. Considering minimum wage in Mexico was 67 pesos daily during my research, this birth package represents 1.3 years of earnings. These births were commonly called *partos humanizados* (humanized birth) by parents, obstetricians, and professional midwives. The women I refer to as "professional midwives" self-identify with this term, are classified as such by their clients, and some (with important exceptions) have undergone formal training and licensure. Their categorization as "professional midwifes" distinguishes them from "traditional midwives" who practice in rural and indigenous contexts (described further, below).

As described by Davis-Floyd (1992), humanized birth is used in Mexico and elsewhere in Latin America to describe birth that purposefully resists medicalization and technocratic practices. The terms "humanized birth," "gentle birth," and "tranquil birth" are used interchangeably to describe a movement aiming to "empower" women and eliminate obstetric violence, and is analogous to the home and water birth movements in the United States, Canada, and Western Europe. Births are often attended by professional midwives (although some obstetricians have joined the movement) and accompanied by doulas.[2] Proponents criticize power inequality inherent in physician-patient relationships and denounce medicalized practices such as unnecessary cesarean sections, episiotomies, isolating birthing mothers in labor and delivery areas, labor induction (including use of hormones such as Pitocin), and repetitive insertion of medical personnel's fingers into vaginas to assess dilation.

My study's participants choosing humanized birth distance themselves from the majority of Mexican society they characterize as entrapped in a political and educational system limiting intellectuality. In line with Currid-Halkett (2017) theory of the aspirational class, these participants value education and New Age bodily practices as mechanisms for accruing cultural capital (Molinié, 2013; Aguilar Ros, 2008). With help of professional midwives representing a foreign or transnational identity, they aspire to join a transnational affluent, cultured community. Many professional midwives in Mexico are either foreign nationals or foreign trained.

Most couples hold advanced degrees in the sciences and humanities (as opposed to technical) fields, and has read widely in English or French. Their commitment to the humanized birth movement is based in their admiration of French obstetricians such as Fréderic Leboyer and Michel Odent, English and American midwifery and "gentle birth" advocates such as Sheila Kitsinger, Ina May Gaskin, Barbara Harper, Thomas Verney, and Robbie Davis-Floyd, and New Age birth methods such as hypnobirthing. Shared passion for humanized birth connects them to global communities in the United States, Canada, England, France, Holland, Spain, Chile, Argentina, and Brazil.

In contrast, the midwife-assisted births of indigenous women in rural contexts were referred to as *partos tradicionales* (traditional births). In referring to births as "humanized" and "traditional," I am using terms I encountered in the field. "Traditional midwives" is the term these practitioners use for themselves; I am not extending an essentialized notion of traditional birth. Findings of my larger ethnographic research project belied the traditional/modern binary (Vega, 2018). However, for many indigenous people across Mexico, certain birthing practices are widely referred to as "traditional." Recently, the Organization of Indigenous Doctors of the State of Chiapas (OMIECH) reiterated the sentiments of many of my indigenous informants when describing traditional birth as being threatened and worth defending.

The traditional births I observed involve indigenous midwives attending indigenous mothers either in the midwife's home-clinic or the birthing mother's home. Since indigenous mothers often arrive with extended family in tow or are surrounded by family members in their home birth environment, these births are sensitive to how complex gender- and generational-hierarchies unfold in specific local contexts. The births often include herbal remedies and massage techniques. Some incorporate biomedical techniques by applying traditional medicine logic to "modern" resources. I observed a traditional midwife apply a postpartum IV to "replace vitamins"—administering IV fluid is cast as a more direct method of providing blood tonic than postpartum soups and teas. Traditional birth is a medically pluralistic form, and, thus, the IV is combined with a postpartum herbal bath. Payment schemes are flexible in quantity, timing, and type, often reflecting the socioeconomic realities in rural indigenous villages.

"Traditional Mexican Midwifery" tourism is unfolding within this context, on both a national and international scale. Tourists are humanized birth proponents, passionate about natural birth and often striving to lead organic lives with an emphasis on holistic health. Their curiosity about "traditional Mexican midwifery" is fueled by a desire to "go back to nature." In their collective imaginary, indigenous birth is not (and should not be) medically pluralistic and does not make use of "modern" resources. Violations to this imaginary can be met with moralistic judgments—one informant openly criticized an indigenous midwife who, in her opinion, was corrupting the traditional art of midwifery by wearing surgical gloves and a sterilized gown while attending birth. From the perspective of humanized birth proponents whom I have interviewed and observed, indigenous birth should be performed as it has been for centuries, thus preserving "Mexican tradition." In this way, indigenous women and their midwives are interpellated as representatives of the precolonial roots of Mexico. Given these divergencies in economic and material resources, moralistic expectations, geographic space, and fictive time, this article uses an intersectional approach to evaluate the effects of "Traditional Mexican Midwifery" tourism on individuals with different positionalities in society.

## Intersectionality and responsible tourism studies

Intersectionality is a multifaceted concept rich with interdisciplinary applications. Here, I explore how the concept has been put to work within tourism studies and adjacent fields of research.

In her recent article (2018), Shelagh Mooney asserts, "The adoption of an intersectional approach by tourism researchers has been relatively limited" (175). She notes a similar underutilization of the concept in hospitality and leisure studies. Likewise, McBride, Hebson, and Holgate (2015) suggest researchers of work and employment relations are not "taking enough notice" of intersectionality as a valuable theoretical tool. Bowleg argues intersectionality is an important theoretical framework for public health, yet "public health studies that reflect intersectionality in their theoretical frameworks, designs, analyses, or interpretations are rare" (Bowleg, 2012, p. 1267). In their respective fields of studies, researchers have tended to study each social category (e.g., race, ethnicity, gender, socioeconomic status, etc.) in isolation. Returning to tourism studies, Mooney advocates an intersectional approach for tourism researchers since exploring one aspect of diversity in isolation is akin to seeing merely one piece of a complex puzzle.

One example of intersectionality being powerfully deployed in tourism studies is Cole's intersectional analysis (2017) of the gender-water-tourism nexus in Labuan Bajo, Indonesia, an emergent tourism destination. Cole, a feminist political ecologist, notes how women bear the brunt of burdens of water scarcity, and examines which women, why, and to what effect. Cole's ethnographic research with over one hundred respondents reveals how patriarchal cultural norms, ethnicity, socioeconomic status, life stage, and proximity to water sources intersect to (re)produce gender inequality. She concludes tourism outcompetes locals for access to water, exacerbating women's suffering.

Identifying a need for more intersectional approaches among tourism researchers, Chambers (2017) suggested "a more critical unpacking of [gender's] ability to produce and (re)produce power relationships in tourism and travel" (as cited in Mooney, 2018). Figueroa-Domecq, Pritchard, Segovia-Pérez, Morgan, and Villacé-Molinero (2015) argue the tourism academy must replace its Western-centric focus with more critical perspectives when exploring race, gender, and other aspects of identity in diverse tourism contexts (as cited in Mooney, 2018).

In direct response to these exhortations, I turn to intersectionality as tool for "decentralising the tourism universe" (Ateljevic, Pritchard, & Morgan, 2007:24-26; also Tucker & Hayes, 2019), framing intersectionality "as a form of social critique so as to foreground its radical capacity to attend to and disrupt oppressive vehicles of power" (Dhamoon, 2011). The ultimate aim of using an intersectional approach is to uncover new insights for creating responsible tourism through social sustainability.

Studies have described responsible tourism as a product consumed by an ethically oriented market segment (Goodwin & Francis, 2003; Weeden, 2013). Grimwood et al. explain, "The term tends to denote a process of planning, policy, and development that prioritizes the community-level involvement, sustainable resource management, equitable distribution of benefits, and minimal negative impacts to local contexts" (Grimwood, Yudina, Muldoon, & Qiu, 2015, pp. 22–23; also Goodwin, 2011; Pickel-Chevalier, Bendesa, & Putra, 2019). Leslie (2012) asserts responsible tourism generates tourism experiences underpinned by holistic, moral concern at multiple levels: the well-being of individual, community, and broader social and physical environment. Criticisms of the practice include it more often serves as a marketing ploy than an ethical planning mechanism (Wheeller, 1991), it aids in the expansion of neoliberalism (Duffy, 2008), and it reproduces power differentials reminiscent of colonial regimes (Sin, 2010). By bringing intersectional theorizing to bear on responsible tourism, my hope is responsible tourism efforts will not merely serve as public relations tactics, but will ameliorate the neocolonial effects of tourism by disrupting existing power inequalities.

Kimberle Crenshaw, reflecting on intersectionality, the concept she first developed, has observed, "It is easier to call for intersectional analysis than perform it" (2011, p. 230; as cited in Mooney, 2018). Crenshaw notes there is not one singular theory of intersectionality, leading to a lack of agreement regarding the intersectional research design process. Additionally, Crenshaw argues we resist moves toward standardization, stating, "intersectional analysis may take us down many roads, but we will only discover what it is by using it" (2011, p. 233). Next, I describe the ethnographic research methods I used for studying intersectionality and "Traditional Mexican Midwifery" tourism.

## Methods

My fieldwork, from October 2010 to November 2013, was conducted as part of my dissertation research. The dissertation was later published as an ethnographic monograph (Vega 2018). My research produced valuable data regarding tourism—specifically, (im)mobilities; the effects of intersectional relationships among gender, class, and race; and lacking sustainability practices—that I have not discussed in publications prior to this article.

Since the founding of anthropology as a discipline, ethnographic fieldwork has been the primary method used for data collection (Malinowski, 1922). Specific methods include in-depth interviewing and participant observation. These generally require anthropologists to live among the population they are studying for an extended period of time. Following in the footsteps of Marcus (1995), Menéndez (1996), Rapp (2000), and Wilson (2004), I used ethnography to create a cartography of midwifery and humanized birth in Mexico, thus identifying different "windows" through which recent shifts in birth practices and health care can be examined (Table 1).

**Table 1.** Setting where data was obtained.

| | |
|---|---|
| Large group settings (workshops, etc.) | 967 |
| Individual or family-level settings | 1,094 |
| Total number of participants | 2,069 |

*Source:* Authors.

I began with professional midwives in Mexico, a contained and connected group of women, subsequently gaining access to their clientele. Concurrently, I interviewed different transnational humanized birth leaders about their respective roles in the movement, both globally and specifically in Mexico. To trace the contours of the cohesive social network emerging around humanized birth, I used a snowball sampling technique to recruit more couples and humanized birth attendants, including physicians and obstetric nurses, to my study (Clatts, Davis, & Atillasoy, 1995). This technique is also called respondent-driven sampling or personal network sampling, since respondents refer the researcher to other members of their personal network.

Throughout my fieldwork, I attended multiple humanized birth conferences and volunteered at two different transnational NGOs, gaining access to training workshops for indigenous traditional midwives. Having befriended a few indigenous midwives, and while staying as a guest in their homes during repeat visits to their villages, I witnessed their interactions with indigenous women and the traditional midwifery care they provide. Finally, I observed medical professionals and maternity patients in both private and public hospital settings and solicited interviews with physicians and policy makers (see Table 2).

This process led me to the States of Guanajuato, Guerrero, Jalisco, México, San Luís Potosí, Veracruz, Chiapas, Oaxaca, Quintana Roo, Morelia, Querétaro, Puebla, Michoacán, and Nuevo León. I traveled to California for interviews, and Brazil for participant observation in" Traditional Mexican Midwifery" tourism (see Table 3). While the geographic breadth of this "field" was enormous, I traveled to meet, observe, and interview specific people who were conscientiously members of a cohesive transnational community. Individuals in my study have acquaintances, and often great friends, among other individuals in my study.

Interviews were semi-structured, lasting from 15 minutes to three hours with the average being approximately forty-five minutes. I tailored my questions to the interviewees' positionality within the humanized birth movement (whether the interviewee[s] was/were a mother, a couple, a humanized birth attendant [professional midwife, obstetrician, obstetric nurse], a traditional midwife, or a policy maker), but usually included questions to help me understand the interviewee(s)' positionality in society (education level, socioeconomic status, ethnicity, etc.). My questions generally followed these themes: his or her occupation, life history, perspectives on gender, the Mexican health system, positive and negative experiences with birth, and the shifting political climate regarding midwifery. By not overstructuring the interviews, I resisted scripting or leading the informants, allowing them to speak for themselves (Briggs, 1986).

Each day, after concluding in-depth interviews and participant observation, I typed up fieldnotes (Emerson, Fretz, & Shaw, 1995). By the end of the 28-month research period, my fieldnotes spanned 734 single-spaced pages. My data analysis is derived from detailed entries in my field diary, audio and video recordings from interviews,

**Table 2.** Types of participants in the study.

| Professional midwives | 111 |
|---|---|
| Physicians | 166 |
| Traditional midwives | 288 |
| Nurses | 65 |
| NGO administrators | 14 |
| Secretary of Health administrators | 17 |
| Hospital administrators | 18 |
| Doulas (birth coaches) | 8 |
| Community health workers | 5 |
| Prenatal educators | 12 |
| Medical school professors | 7 |
| Professional midwifery school instructors | 22 |
| medical anthropologists | 9 |
| Public health/public policy professionals | 11 |
| Mothers | 125 |
| Fathers | 31 |

*Source:* Authors.

**Table 3.** Geographic distribution of participants in the study.

| Guanajuato | 306 |
|---|---|
| Guerrero | 137 |
| Jalisco | 247 |
| Federal District | 68 |
| San Luís Potosí | 296 |
| Veracruz | 700 |
| Chiapas | 85 |
| Oaxaca | 72 |
| Quintana Roo | 103 |
| Michoacán | 8 |
| Nuevo León | 18 |
| California (individuals who work in Mexico) | 7 |
| Mexico *and* the United States | 20 |
| Multiple Mexican states | 3 |
| Elsewhere | 40 |

*Source:* Authors.

and digital ethnographic data gleaned from social media. Upon concluding my research, I engaged in an iterative process, using open coding to identify emergent themes and synthesize higher order constructs. The primary themes emerging from this open coding process included the accrual of cultural capital; gender and class inequality; racialization processes; social mobility; and obstetric violence. Subthemes included the commodification of indigenous culture, ethnomedical tourism, unequal access to citizenship-based resources, and aspirational strategies within a meritocracy.

Open coding is important because it ensures data-driven analysis. Instead of using data analysis software to search my fieldnotes for *repeated terms*, I carefully read through my 700-plus pages of fieldnotes and created notecards for each *emergent theme*. While this hand coding was labor intensive, I recognized intersectional linkages between class, gender, and race that qualitative data analysis software would have missed.

All names throughout this ethnography are pseudonyms, except for humanized birth public figures whom I encountered at humanized birth events but were not direct participants in my research. I obtained IRB approval for this research from the University of California, Berkeley.

**Table 4.** Women who were pregnant or recently pregnant in the study.

| | |
|---|---|
| Pregnant women | 19 |
| Women in active labor | 17 |
| Births (witnessed as part of the research) | 7 |
| Postpartum | 33 |

*Source:* Authors.

## *Hyper-self-reflexivity*

The ethnographic vignettes appearing in this article are "thick descriptions" synthesized from fieldnotes written immediately after concluding participant observation each day (Geertz, 1973). As such, the research methodology is ethnography, not autoethnography. I chose to write from a first person perspective to be transparent about my positionality with respect to the research (see Crossley, 2019). Since ethnographers use their senses as their primary tools for data collection, all ethnography is, by definition, data that has been perceived through the ethnographer's eyes and ears. I was not the subject of my research (See Ellis & Bochner, 2000); instead, I used hyper-self-reflexivity to tell a story about study participants.

I was methodologically and ethically committed to Spivakian hyper-self-reflexivity (Kapoor, 2004)—that is, acknowledging and intentionally resisting my own privilege to gain greater access to the lived experiences of my informants. This commitment was important given this article's critique of multivalent processes leading to the inadvertent reinscription of social inequality between disadvantaged and privileged informants. While I am a woman of "ethnic" heritage (Mexican, Nahua, Blackfoot, and Chinese descent), I am also highly educated, privileged, and reared in urban and suburban United States. I cannot claim to be a subaltern, "native" informant. During my research, I "unlearned my own privilege" (Spivak, 1988, p. 287; Spivak & Harasym, 1990, p. 9) by adjusting to the socioeconomic standard of my research contexts; i.e., limiting my monthly budget to five thousand Mexican pesos (at the time, approximately four hundred dollars). Despite these efforts, I recognize "shedding privilege" can never be fully achieved.

My work uncovered "partial connections" (De la Cadena, 2015) and brought multiple "situated knowledges" under the same lens. It was sensitive to Haraway's concern about inadvertently romanticizing and appropriating while claiming to see from positions of the less powerful (Haraway, 2014:45). I readily recognized my "partial perspective" (Haraway, 2014) and admitted my potential as an ethnographer was limited.

Relying on translators limited my access and substantially changed the dynamic of interviews. I placed women's comfort and privacy concerns during intimate moments over my desire for ethnographic material. This meant occasionally choosing not to be present at the moment of birth (see Table 4). While I have extensive training in the health sciences, I am not a midwife or medical professional with technical expertise; thus, I have never intervened during births.

Practicing hyper-self-reflexivity has been advantageous. I was able to interview informants and conduct participant observation at both ends of the socioeconomic spectrum (Nader, 1972). Using hyper-self-reflexivity, I resisted *speaking for* the indigenous "Other" or placing the humanized birth tourist at the center. Postcolonial studies can sometimes fail to include the communities they seek to serve, thus reinscribing power differentials between the colonized and the colonizer (Grimwood et al., 2015). Resisting a Global North-centric perspective and instead striving to promote a multi-perspectival approach, I built upon Yan and Santos' (2009) critique of how much of the literature on tourism focuses on Western perceptions of the Other, thus missing how the Other represents itself in contemporary tourism discourse. My hyper-self-reflexive research method avoided positivist assumptions of universality and generalizability, and, therefore, augmented an intersectional approach to tourism studies.

### *Intersectional axes of analysis: gender, race, and class*

I have selected three intersectional axes of analysis: gender, race, and class. Gender is an important axis since the women in my study—including those who give birth via "humanized" and "traditional" methods—are pursuing alternatives to a biomedical system that often wrests power away from women and places it in the authoritative hands of (often male) obstetricians. While the concept of intersectionality helps to describe the sameness characterizing them, it also sheds light on the differences in their experiences. Their divergent positionalities in society are based on race and class. They seek alternatives to medicalized birth for different reasons, with different access to resources, and, ultimately, with different outcomes. By recognizing intergroup differences in groups presumed to be homogenous (in this case, "women"), the theoretical framework of this article dismantles positivist assumptions of universality and generalizability (Mooney, 2018).

I merge Crenshaw's concept of "intersectionality" with Bridges' perspective on pregnancy as a "site of racialization" and "race as a process." I draw attention to the ongoing, interfacing processes leading to the material (re)realization of racial inequality.

My perspective differs from Bridges since for her, in order to "get to" questions of race, she first had to "go through" issues of class: "The analytic of race had to be folded into an analytic that began with class" (Bridges, 2011:9). The experiences I encountered imbued in me an understanding of racialization as a process through which multiple axes of inequality intersect simultaneously.

When people first meet, they engage in an immediate, predominantly unconscious, mental calculus that includes race *and* class, and ultimately maps race *onto* class, *and vice versa*. This mental calculus incorporates phenotype, dress, education, language, nationality, etc. into racialized perceptions. Through ethnographic examples, I demonstrate how one's intersectional positionality in society influences one's perception of oneself and others, and, in turn, how one is perceived by others.

Departing from Bridges, the multivalent ways in which women across Mexico use pregnancy as a site of racialization leads me to conclude the physical body is only one of many complexly related signs of racial difference, albeit an important one. The commodification of indigenous culture, differential access to citizenship-based resources, disparate degrees of social and physical mobility, and race- and class-specific expectations for motherhood is also windows for examining how intersectional racialization processes are woven into the everyday fabric of social life.

## Romanticized visions of indigenous "others" (and the possibility of self-romanticization)

It became clear during the early stages of my research that one of the primary racializing effects of "Traditional Mexican Midwifery" tourism unfolded through romanticized visions of indigenous "Others." Over time, my ethnographic observations revealed some indigenous study participants leveraged these racializing discourses to their own benefit. A few entries from my field diary will illustrate these points.

> *I boarded a second-class bus to Matehuala, connecting to Estación Catorce where I attended an event billed as an encuentro (a meeting of two cultures) between professional medical personnel and traditional midwives. It was sponsored by Secretary of Health's office for Traditional Medicine and Intercultural Development (TMID) and organized by an anthropologist.*
>
> *Nurses, physicians, anthropologists, and professional midwives, emerging as* the *source of expert knowledge, gave PowerPoint presentations to traditional midwives. Many were interested in New Age therapies; some even dedicated their presentations to the benefits of medicinal plants. While watching various professionals teach "traditional" remedies to traditional midwives, I perused the agenda for the three-day event. Will traditional midwives have an opportunity to share* their *expert knowledge? The last two entries, at the end of the final day, were presentations by "traditional midwives." While all other entries named the presenting professional and presentation title, "traditional midwives" were unnamed and no information was given regarding their presentations.*
>
> *Scanning the crowd, I noticed piercings and tattoos, foreigners with camping gear, and many people from Mexico City. Beside the registration table was a display of natural products for sale (shampoos, teas, marmalades, homeopathic remedies, etc.) These products and event publicity materials were branded with a cosmic/floral image and the Nahuatl word "Nanahtli." Staffing the table was Sofia Gil, friend of the anthropologist who organized the event. Traditional midwives were not involved in producing these items; however, one traditional midwife, Liliana López, was tasked with manning the table for a few hours—her expressed boredom and disinterest made me wonder about her feeling of dispossession.*
>
> *Curious about the disjuncture between the encuentro's promotion and its manifest reality, I traveled to the TMID office in Mexico City to interview two planners of government health programs about the relationship between their published work on "interculturality" and what had unfolded at the encuentro. Explaining they sponsored the encuentro but were not involved in its planning or execution, both doctors assured me what I had witnessed is not representative of their work. They have planned and executed more than fifty highly successful Encounters for Mutual Enrichment Between Health Personnel and Traditional Midwives at which traditional midwifery knowledge was "rescued" from the ongoing threat of extinction. When I asked about their next encuentro, stating I would travel any distance to attend the type of event they described as it would be very informative for my research ... . They advised Encounters for Mutual Enrichment are no longer being planned.*

The encuentro is an example of economic and political inequalities between indigenous and nonindigenous peoples of Mexico, demonstrating how political economic power determines tourism development. It illustrates how indigenous midwives—nameless and interchangeable on the conference agenda—are excluded from determining their cultural images or representations, deciding how their midwifery and traditional medicine practices will be commodified and sold, and benefiting from the economic profits. Most of the literature on tourism as neocolonialism focus exclusively on international tourism. While this example involves mixed international and national tourism, it highlights ongoing exploitation of indigenous people and their cultural practices. This article critiques how indigenous midwifery practices are usurped, while indigenous midwives are excluded from decision-making processes and profit streams relating to the development of "Traditional Mexican Midwifery" tourism.

My ethnographic observations in Mexico suggest "Traditional Mexican Midwifery" tourism aims to present a precolonial fantasy of indigenous practices as "pure," "noble," and "untouched" by colonialism and modernity. Indigenous midwifery is described by nonindigenous humanized birth proponents as having maintained a primordial connection with nature through the preservation of centuries-old traditions. This fantasy is a "daydream" based on stereotypes and myths (Said, 1978:52). I argue the stereotyped images of indigenous midwives currently being presented for consumption by popular Mexican midwifery tourists in fact represent the attitudes and opinions of the (neocolonial) tourists themselves (Kabbani, 1986; Volkman, 1990; and Picard, 1991). That is, "Traditional Mexican Midwifery" tourism confirms the myths and images already held by tourists about indigenous "Others." "Traditional Mexican Midwifery" tourism "reinforce[s] the myths of the past through the symbols of the present" (Palmer, 1994:34). While it purportedly introduces tourists to "indigenous practices," it primarily reflects globalized humanized birth principles, seen through a New Age, deauthenticated, reinvented "traditional" lens. Stated simply, indigenous traditions are remade "anew" through New Age iterations of humanized birth.

My ethnographic observations of "Traditional Mexican Midwifery" tourism demonstrate how indigenous practices are converted into a tourist attraction. "Traditional Mexican Midwifery" tourists are drawn to the allure of "ancient traditions." This allure, however, is fabricated through neoliberal marketing. Indigenous people are not caught in a time warp, frozen in a past time, or live their lives and carry out their practices in a social vacuum. Humanized birth practitioners and traditional birth attendants come into frequent contact during training workshops, conferences, and fairs. While these encounters fuel the fetishization of indigenous practices by nonindigenous individuals, it has translated into material benefits for only a few specific indigenous participants, and has inadvertently reinforced racial hierarchies rooted in colonial legacies.

While the overt prejudice and racial discrimination of the colonial era no longer exist, neocolonial tourism in the form of "Traditional Mexican Midwifery" tourism subtly reinforces romanticized stereotypes of the indigenous "Other," thereby reinscribing hegemonic structures rooted in the colonial past. I am not arguing "Traditional Mexican Midwifery" tourism creates racism—it is clear racism existed centuries before (Farrell, 1979; Manning, 1979; La Flamme, 1979)—however, my ethnographic research suggests this type of tourism reinforces preexisting structures of intersectional inequality, including racializing processes. "Traditional Mexican Midwifery" tourism is, thus, "neocolonial" in the sense it perpetuates inequality. By excluding in practice those it romanticizes, this type of tourism inadvertently asserts the superiority of one group over another.

Continuing my effort to "decentralize the tourism universe"— I turn to ethnographic observations with indigenous informants to exemplify what would likely happen if indigenous individuals were incorporated as agents in "Traditional Mexican Midwifery" tourism. Given the opportunity, some indigenous individuals choose to leverage their own indigeneity as a source of cultural capital within humanized birth circles. Certain indigenous participants in my study used their indigeneity as a source of cultural capital, which in turn opened opportunities for increased social and economic capital. These indigenous informants realized by playing into the indigenous identity nonindigenous consumers imagine and fetishize, they could travel, gain renown in transnational humanized birth circles, and earn a living otherwise inaccessible to them. Two examples of this are Yanira García and Eugenia Navarro, traditional midwives in Tepotzlán and the High Mountains of Veracruz, respectively.

*I traveled to Tepoztlán—an artsy community of affluent Mexicans and American expatriates. to visit with Yanira García. While this was our first meeting, I already knew something about the traditional midwife from an American-made documentary on her work. When the documentary was filmed, Yanira was attending birthing mothers at a feverish pace; both Mexican and foreign women traveled to Cuernavaca to give birth in her thriving home-clinic.*

*Yanira explained how she became a midwife and the unique position she holds within the transnational humanized birth community. Her mother, two aunts, grandmother, and two great-grandmothers were all midwives and* curanderas. *Yanira was first introduced to the transnational humanized birth community when she met Jan Tritten, editor-in-chief of Midwifery Today, at a midwifery conference. She explained Midwifery Today invites "the most traditional" midwives to speak internationally, all expenses paid. As her reputation and her birthing clinic grew, Yanira began receiving apprentices from Austria, Denmark, Norway, Italy, Spain, Brazil, Costa Rica, the United States, and Israel. She organizes "exchanges" with these apprentices and has traveled to their countries to speak about her work.*

*Yanira is unique since she, and a few others across Mexico, has constructed an identity employing place-based knowledge as currency within transnational circles. Yanira's celebrity is the result of an inherent contradiction: despite becoming transnational, she continues to be perceived as mononational. Yanira has become a traveling representative of the "unchanging" indigenous roots of Mexican birthing traditions. I wondered if Yanira's paradoxical position is engendered by the transnational humanized birth community's fetishization of indigeneity—a fascination allowing for a select few indigenous women to serve as representatives of traditional Mexican midwifery at international conferences, but does not permit merging of "traditional" and "professional" midwifery.*

*Yanira's case piqued my interest since she was selected for being one of the "most traditional" midwives in Mexico; however, she does not speak Nahuatl, and practices popularized "traditional" methods. Throughout my research, I had many opportunities to observe midwives who primarily or exclusively speak indigenous languages and whose repertoires comprise of remedies and techniques unknown in transnational circles; yet these midwives are not present at humanized birth conferences or Mexican Midwifery Association meetings.*

*In contrast, Eugenia is a primarily Nahua-speaking midwife who attends an exclusively Nahua clientele. Over several years, we formed a friendship through visits, letters, and phone calls. After many intimate conversations, I asked Eugenia to sit for a formal, video-recorded interview, explaining this footage could potentially be edited into an ethnographic film used to demonstrate the work of traditional midwives to American anthropology students and conference attendees. She acquiesced, on one condition: that I not start the video camera until she had finished dressing herself in her traditional indigenous attire, put on her best jewelry, and combed her hair. I agreed. She searched among several plastic bags, finally selecting the traditional blouse she wanted to wear while being filmed. She folded the pleats in her bata (a large piece of black wool cloth worn as a skirt) and straightened the lace and ribbons on her blouse in a methodical, almost ritualistic fashion. When I asked Eugenia for a formal interview, she exuded a sense of pride as she performed her "indigeneity" for imagined spectators. I was struck by the politics of representation and how different informants make concerted efforts to portray themselves in ways that meet disparate goals.*

In the prior field diary entry, Eugenia carefully selects her clothing based on how she hopes to be perceived by others. This behavior is not unique to her or her context. Everyone engages in similar behavior. Readers of this article have dressed in an intentional manner for a job interview, and the attire selected likely differing from their everyday contexts and with the goal of giving an impression of professionalism. What separates Doña Eugenia is the image she hopes to create is one of romanticized indigeneity. While the field diary entry from Estación Catorce demonstrates how non-indigenous study participants romanticized indigenous "Others," Doña Eugenia's actions show how indigenous participants at times engage in self-romanticization, thereby leveraging their indigeneity as a source of cultural capital.

This example demonstrates that indigenous midwives are strategic individuals not totally lacking agency; however, I observed significant limitations to the agency they possess. A strength of intersectional approaches is they facilitate analysis beyond binary determinations. The same traditional midwives who successfully converted their indigeneity into a source of cultural capital faced obstacles when attempting to convert their cultural capital into professional recognition. I argue these obstacles are rooted in socioeconomic inequality.

To illuminate this point: Mexican Midwifery Association has a section focusing on traditional midwives, thereby demonstrating interest in traditional midwives as *objects*. Concurrently, traditional midwives are barred from the Association's professional guild. While many of the women heading the Association are not licensed to practice midwifery in Mexico—some are trained in the United States, others have earned certificates through online programs, while still others are trained in allied professions like obstetric nursing—they are all eligible for the protections and benefits of the professional guild based on having obtained "some form of academic training." This criterion for inclusion is so inexact that, I argue, it primarily serves as a form of gatekeeping—an obstacle that "uneducated" (read: lower-class) midwives cannot afford to overcome.

## "Traditional Mexican Midwifery" tourism: the offshoring of neocolonial tourism?

Williams argues the tourism sector, "in its very nature, has perpetuated the legacy of colonialism, so much so that it may be conceptualized as a 'neo-colonial phenomenon'" (Williams, 2012:191). Through continued colonial servitude, tourism transfers power from the destination country to foreign corporations (Williams, 2012:191; see also Gmelch, 2003; Zhang, Lee, & Xiong, 2019; and Wong, 2015). My participant observation in Mexican tourist destinations revealed that locals, surrounded by luxuries they can never partake in except as service workers, are painfully aware of their inferior status in the socioeconomic order (Vega, 2018).

Using the lens of" Traditional Mexican Midwifery" tourism, my ethnographic research explores tourism capitalizing on interest in traditional Mexican midwifery while excluding indigenous midwives (and representatives of Mexico more generally). Not only has political and economic power been wrested away from indigenous individuals, their visibility and ability to leverage their indigenous identities for increased cultural capital have also vanished. Stated differently, not only does this type of tourism fail to meet the needs of the indigenous people it fetishizes (from an extended distance), it actually excludes them from the equation entirely.

In the case of "offshoring" neocolonial tourism, indigenous individuals do not even have the admittedly problematic option of portraying themselves as the "objects" tourists wish to gaze upon (Urry, 1990:48). When "Traditional Mexican Midwifery" tourism is exported from Mexico, indigenous people have been physically removed from the touristic representations of their practices. In their absence, middle- and upper-class transnationals control how indigenous practices are represented for touristic consumption. What is consumed by popular Mexican midwifery tourists is not indigenous Mexico, but, rather, a vision of indigenous Mexico (Clifford, 1988:55). While a concern regarding global tourism is it integrates tourist destinations into the global capitalist system, when "Traditional Mexican Midwifery" is offshored, a vision of indigeneity is consumed while the indigenous individuals themselves are excluded from capitalist streams of commerce. A field note from Brazil will demonstrate this point.

Adeli Hirsch is a professional midwife renowned in the humanized birth community. I attended her international workshop on "traditional Mexican midwifery" in Florianópolis, the capital city of Santa Catarina, Brazil. The city, known for its high quality of life, unparalleled Human Development Index score among Brazilian capitals, nightlife, tourism, and dynamism, is second home for many Argentines, North Americans, Europeans, and Paulistanos; as a result, it is perhaps the "whitest" city in Brazil.

*Adeli, reared in the United States and Mexico, is of Jewish and Mexican heritage. Like other successful professional midwives in Mexico, she studied midwifery in Texas. Most of the participants had traveled to Brazil from other Latin American countries to learn from Adeli. I was surprised how fair the group was—although we were in Brazil, there were very few women of African descent. As we introduced ourselves, Adeli commented I was the only Mexican at this workshop on "Traditional Mexican Midwifery." I quickly explained that I, a person of mixed ethnic heritage reared in the United States, am an inadequate representative of Mexico.*

*For most of the week, we sat on the beach or in boats, listening to Adeli's anecdotes of births she attended and her personal reproductive experiences. As the workshop transpired, I realized very little of what was taught is traditional Mexican midwifery. Most of the techniques Adeli discussed are not the traditional techniques of indigenous midwives in Mexico. Her workshop included New Age explanations of homeopathy and how to make tinctures from placentas and placenta art. I increasingly wondered if what I was observing can be more aptly described as" Traditional Mexican Midwifery" tourism: middle- and upper-class women from across Latin America who are not midwives but interested enough in New Age notions of traditional midwifery, indigeneity, and "going back to nature" to travel internationally and spend a week on Brazilian beaches with someone offering herself as a representative of traditional midwifery knowledge.*

*My curiosity of whether the workshop was an example of traditional midwifery popular tourism became more persistent after we participated in a nighttime* temascal *ritual. Compared to* temascales *prepared by indigenous people in Mexico for healing purposes, this* temascal *was decidedly mild. The temperature was not as extreme; the door was opened frequently for participants to cool off; and there was no flogging with fragrant herbs.*

*After the ritual, the women emerged, crying, hugging, and kissing each other. As we shared our experiences, most of the women recounted the* temascal *had forced them to deal with old emotional traumas, and by the end of the ritual many had let go of fears and pain they had been harboring.*

What struck me about this experience was how the temascal ritual had been extracted from its original geographic and sociocultural contexts; usurped, transported, and manipulated for profit within Florianópolis' tourism industry; and infused with New Age meanings which led to experiences of emotional cleansing and psychological healing among the participants that were wholly distinct from the way my indigenous informants experience the temascal. How does the New Age notion of healing allow for the mixing of concrete practices originating from disparate contexts and rooted in divergent ideologies? What are the unintended consequences of this ostensibly clean extraction of healing practices from the social milieu for which they were created?

The techniques Adeli taught were not authentic representations of traditional Mexican midwifery. From my extensive participant observations with both indigenous midwives and humanized birth practitioners, I recognize these techniques as a New Age reimagination of what indigenous birthing techniques might be like from the perspective of nonindigenous transnationals. Nonetheless, the workshop was premised on Adeli being a representative of traditional indigenous knowledge. The fact that professional midwives go unquestioned when they stake claims to traditional knowledge, while traditional midwives are excluded from a national guild for professional midwives unless they undergo a formal course of study, speaks to the unequal power relations operating within Mexican midwifery.

This fieldnote in Brazil exemplifies how "Traditional Mexican Midwifery" tourism has been offshored—appropriating indigenous midwifery practices while fully excluding indigenous people. "Traditional Mexican Midwifery" tourism is thus an example of how "cultural heritage has been commoditized in a systemic way.... Tourism increasingly appropriates spaces and redefines the meaning of cultural objects" (Cywinski, 2015:21). "Traditional Mexican Midwifery" tourism has redefined the meanings of indigenous practices, thus maintaining an "ideological barrier" with colonial roots.

At first glance, the humanized birth movement among Latin American women appears to be an effort to move beyond the textual politics of science and technology to something even more radical. Through alternative and "traditional" birth techniques, humanized birth proponents aim to take back the power of women through ownership of their bodies' natural ability to give birth, thereby "seizing the tools to mark the world that marked them as other" [Haraway, 1990:175]). While this would seem like a cyborg project of the twenty-first century, humanized birth does not satisfy Haraway's utopic vision since it is not "rooted in claims about fundamental changes in the nature of class, race, and gender" (Haraway, 1990:161).

Haraway envisions "an emerging system of world order analogous in its novelty and scope to that created by industrial capitalism" (Haraway, 1990:161). In contrast, the humanized birth movement has given rise to global tourism practices deeply embedded in industrial capitalism. Haraway points to how science and technology provide fresh sources of power. My ethnographic fieldnotes demonstrate that although it purports to, "Traditional Mexican Midwifery" tourism does not invert this power. While the participants I describe are all women, those who control the representation of cultural products for tourists' consumption and those who are excluded from participating are categorically distinct based on class and race. In this way, "Traditional Mexican Midwifery" tourism exemplifies intersectional inequality.

For these reasons, I argue New Age, "traditional," and "holistic" therapies are not a panacea for the biomedical control of women's bodies. While they may be beneficial and liberating for some women, they exclude other women by reinscribing colonially-rooted inequalities. As with biomedical techniques, these therapies also require fresh sources of analysis and political action (Latour, 1987).

## Conclusion: Seeking social sustainability in "Traditional Mexican Midwifery" tourism through responsible tourism?

Tourism can be an indigenous process (in which local residents both motivate and reap the benefits of tourism), or it can serve a neocolonial agenda (in which tourism development and profits unequitably favor transnational elites). This distinction illustrates how political economic inequality intersects with racialization processes during tourism development. In the case of "Traditional Mexican Midwifery" tourism, I argue romanticization of indigeneity without including the actual views and needs of indigenous individuals is a clear indication of neocolonialism. "Traditional Mexican Midwifery" tourism does not rely primarily on indigenous midwives to determine how their midwifery practices will be represented and circulated on the global stage.; instead, "indigeneity," "traditionality," and "nature" are conflated and subsequently commodified/fetishized by nonindigenous "allies" for economic profit. The ethnographic portions of this article have described inadvertent consequences when cultural tourism and medical tourism merge without including those whose ethnomedical practices are being consumed.

In conclusion, I explore the relationship between neocolonial tourism and social sustainability, adopting Smith, Lommerse, and Metcalfe (2014) tripartite model of social sustainability. Smith et al. (2014), writing from the perspective of interior architecture, emphasize architects' responsibility to work collaboratively with communities. Architects must strive to meet the needs of local residents, be agents of social justice, and protect cultural heritage. Based on this definition, I argue neocolonial forms of tourism such as "Traditional Mexican Midwifery" tourism are not socially sustainable. It fails to meet the needs of the indigenous people it purportedly represents when it removes indigenous individuals from profit streams and deauthenticates indigenous practices. It belies the very essence of social justice and cultural heritage.

To suggest a more sustainable alternative, I bring existing literature on responsible tourism to bear on ethnomedical tourism such as" Traditional Mexican Midwifery" tourism. To my knowledge, this article is the first to bring together debates regarding responsible tourism and medical tourism. While this article focuses on the specific example of "Traditional Mexican Midwifery" tourism, the lens of responsible tourism can be applied to a range of medical tourism practices, and "ethnic" tourism practices more generally. These novel applications of the responsible tourism concept points to its potential for (re)defining social sustainability. Thus, I am borrowing the concept of responsible tourism and reconfiguring it to emphasize what it offers with respect to the *cultural* aspects of sustainability.

I argue "Traditional Mexican Midwifery" tourism is a product consumed by a market segment that is "ethically oriented." Humanized birth proponents care deeply about women's empowerment, reducing unnecessary cesarean sections, and eliminating obstetric violence. My use of the responsible tourism framework highlights the good intentions of humanized birth proponents while also providing suggestions how their tourism practices can be improved for greater equity and inclusion. Grimwood et al. write, "Even good intentions discipline us to ignore certain truths, are never without their silences or modes of othering, and are always ripe for critical dialogue and debate" (Grimwood et al., 2015, p. 23). This statement most definitely applies to humanized birth discourse. Grimwood et al. clarify their critique should not be construed as a denunciation. They highlight the potential virtue and value of responsible tourism and argue it should not be jettisoned altogether. They write, "Power circulates within discourses of responsibility to normalize particular versions of 'truth,' dismiss the presence of others, and reinforce certain kinds of social privilege and disenfranchisement. When we talk and write about responsibility in tourism, we invariably represent specific interests, values, and power relations" (Grimwood et al., 2015, p. 35).

The unequal politics of representation and reentrenchment of intersectional inequalities I observed in "Traditional Mexican Midwifery" tourism are examples of what Grimwood et al. critique as the misapplication of responsible tourism. I applaud the value of humanized birth practices, and the virtue of eliminating violence toward women while empowering them to seize back control over their bodies and births. By no means do I wish to undermine the important work being accomplished by the humanized birth movement and its proponents. However, it is also important to note humanized birth discourse represents the interests and desires of its majority-affluent participants, inadvertently reinforcing their privilege and reinscribing the disenfranchisement of "unspeaking" indigenous "Others."

Research on "Traditional Mexican Midwifery" tourism clarifies the apparent contradiction between Molinié's argument and Aguilar Ros' (2008) conceptualization of the New Age as a postmodern, cosmopolitan, and countercultural alternative to materialism and consumption. I observed how desire among the "global professional class" (Spivak, 2003) for "going back to nature" and "traditional ways of birthing" reinforces socioeconomic stratification and processes of racialization in the name of humanization and commodifies culture under the guise of rejecting consumption. The great irony is these parents view themselves as actively resisting materialism, but their very refusal of the consumerism, nationalism, and late capitalism they diagnose leads to the commodification of culture, heightened moralization and gradation of motherhood, reinscription of racial hierarchies, and (false) appropriation of indigenous notions of spirituality. The humanized birth movement explicitly avoids direct references to consumption. These mothers reject expensive baby carriages, nanny services, and similar markers of commodified or elite motherhood. They seek cloth wraps to bind their children to their bodies and promote mother-child bonding. Their cloth wraps, however, are not woven shawls indigenous women use; they are sold by companies using the internet to market humanized birth products as noncommodity commodities. "Traditional Mexican Midwifery" tourism serves as another instantiation of a noncommodity commodity.

Dismantling intersectional inequality in "Traditional Mexican Midwifery" tourism requires coordinated initiatives among Mexican Midwifery Association, Mexican Secretary of Health's Traditional Medicine and Intercultural Development (TMID), and critical tourism scholars and activists.

To unseat class- and race-based inequality, Mexican Midwifery Association must cease viewing indigenous midwives as an object of interest and instead engage them as equals in the Association. This would mean redefining "professional training" within the Association so the empirical knowledge of indigenous midwives is valued and legitimized. Indigeous midwives should be welcomed into the professional guild and granted the same rights and protections as their academically-trained counterparts.

While Mexican Secretary of Health's TMID sponsored the Estación Catorce *encuentro*, it did not fulfill its promise of mutual exchange between biomedically-trained health professionals and indigenous midwives. TMID must hold event organizers accountable for fulfilling the office's stated purpose of supporting traditional medicine and intercultural development. Better still, TMID should return to planning and executing Encounters for Mutual Enrichment Between Health Personnel and Traditional Midwives and, in the process, directly confirm these encounters are characterized by mutual exchange.

Lastly, critical tourism scholars and activists play a valuable role when it comes to realizing the potential of responsible tourism through heightened social sustainability. Critical tourism scholars should conduct further community-engaged research on ethnomedical tourism using an intersectional approach. To effect positive changes in "Traditional Mexican Midwifery" tourism, critical tourism activists are needed to sensitize Mexican Midwifery Association and Secretary of Health's TMID to hidden cultural commodification. Effecting change at the grassroots level begins with spreading awareness about how celebration of "traditional" ways of birthing may inadvertently commodify indigeneity (Vega, 2017a) while excluding indigenous people from humanized birth discourse, cultural images and representations, and profit streams. Only through heightened awareness can intersectional inequalities in "Traditional Mexican Midwifery" tourism begin to come undone, thereby opening the door for responsible tourism that accomplishes its goal through social sustainability.

## Notes

1. For a discussion of different analytical levels at which social divisions need to be studied, their ontological base, and their relations to each other, see Yuval-Davis, 2006.

2. Doulas are individuals—generally women—who are trained as birth coaches. Their primary role is to accompany, encourage, and comfort mothers during active labor. They also serve as representatives of the birthing mother's interest by advocating for her birth plan.

## Acknowledgements

The research for this paper was conducted with fellowship support from the Jacob K. Javits Foundation, the Regents of the University of California, and the Ford Foundation.

## Disclosure statement

No potential conflict of interest was reported by the author(s).

## References

Aguilar Ros, A. (2008). Danzando a Apaxuki: La Semana Santa en San Andrés Cohamiata desde los mestizos visitantes. In K. Argyriadis, R. de la Torre, C. Gutiérezz Zúñiga, & A. Aguilar Ros (Eds.) *Raíces en movimiento. Prácticas religiosas tradicionales en contextos translocales* (pp. 59–192). Guadalajara, MX: El Colegio de Jalisco/CEMCA/CIESAS/IRD/ITESO.

Ateljevic, I., Pritchard, A., & Morgan, N. (2007). *The Critical Turn in Tourism Studies: Innovative Research Methods*. Abingdon, UK: Taylor & Francis.

Bowleg, L. (2012). The problem with the phrase women and minorities: Intersectionality-an important theoretical framework for public health. *American Journal of Public Health*, *102*(7), 1267–1273. doi:10.2105/AJPH.2012.300750

Bridges, K. M. (2011). *Reproducing Race: An Ethnography of Pregnancy as a Site of Racialization*. Berkeley, CA: University of California.

Briggs, C. (1986). *Learning How to Ask: A Sociolinguistic Appraisal of the Role of the Interview in Social Science Research*. Cambridge, UK: Cambridge University Press.

Chambers, D. (2017). Are we all in this together? Gender intersectionality in tourism. *Keynote address at the Critical Tourism Studies Conference VII 2017*. Majorca, Spain: University de les Iles Balears.

Clatts, M. C., Davis, W. R., & Atillasoy, A. (1995). Hitting a moving target: The use of ethnographic methods in the development of sampling strategies for the evaluation of AIDS outreach programs for homeless youth in New York City. In E. Y Lambert, R. S. Ashery, & R. H. Needle (Eds.), *Qualitative methods in drug abuse and HIV research, NIDA research monograph 157* (pp. 117–135). Rockville, MD: National Institute on Drug Abuse.

Clifford, J. (1988). *The predicament of culture*. Boston, MA: Harvard University.

Cohen, C. J. (1997). Punks, Bulldaggers, and Welfare Queens: The Radical Potential of Queer Politics. *Glq: A Journal of Lesbian and Gay Studies*, *3*(4), 437–465. doi:10.1215/10642684-3-4-437

Cole, S. (2017). Water worries: An intersectional feminist political ecology of tourism and water in Labuan Bajo, Indonesia. *Annals of Tourism Research*, *67*, 14–24. doi:10.1016/j.annals.2017.07.018

Crenshaw, K. W. (2014). The structural and political dimensions of intersectional oppresion. In P. R. Granzka (Ed.), *Intersectionality: A foundations and frontiers reader* (pp. 16–21). Boulder, CO: Westview.

Crenshaw, K. W., Gotanda, N., Peller, G., & Thomas, K. (1995). *Critical race theory: The key writings that formed the movement*. New York, NY: New Press

Crossley, É. (2019). Deep reflexivity in tourism research. *Tourism Geographies*, 1–22. doi:10.1080/14616688.2019.1571098

Currid-Halkett, E. (2017). *The sum of small things: A theory of the aspirational class*. Princeton, NJ: Princeton University.

Cywinski, P. (2015). Tourist Neo-colonialism as an Indication of the Future of Islands. *The Example of Borobodur (Central Java). Miscellanea Geographica—Regional Studies on Development*, *19*(2), 21–24.

Davis-Floyd, R. (1992). *Birth as an American rite of passage*. Berkeley, CA: University of California.

De Genova, N. (2005). *Working the boundaries: Race, space, and "llegality" in Mexican Chicago*. Durham, NC: Duke University.

De la Cadena, M. (2015). *Earth beings: Ecologies of practice across Andean worlds*. Durham, NC: Duke University.

Dhamoon, R. K. (2011). Considerations on mainstreaming intersectionality. *Political Research Quarterly*, *64*(1), 230–243. doi:10.1177/1065912910379227

Duffy, R. (2008). Neoliberalizing nature: Global networks and ecotourism development in Madagascar. *Journal of Sustainable Tourism*, *16*(3), 327–344. doi:10.1080/09669580802154124

Ellis, C., & Bochner, A. (2000). Autoethnography, personal narrative, reflexivity: Researcher as subject. In N. Denzin & Y. Lincoln (Eds.), *Handbook of qualitative research* (2nd ed., pp. 733–768). Thousand Oaks, CA: SAGE Publications, Inc.

Emerson, R. M., Fretz, R. I., & Shaw, L. L. (1995). *Writing ethnographic field notes*. Chicago, IL: University of Chicago Press.

Farrell, B. (1979). Tourism's human conflicts: Cases from the pacific. *Annals of Tourism Research*, *6*(2), 122–136. doi:10.1016/0160-7383(79)90143-9

Figueroa-Domecq, C., Pritchard, A., Segovia-Pérez, M., Morgan, N., & Villacé-Molinero, T. (2015). Tourism gender research: A critical accounting. *Annals of Tourism Research*, *52*, 87–103. doi:10.1016/j.annals.2015.02.001

Foucault, M. (1990 [1978]). *The history of sexuality*. UK: Vintage.

Geertz, C. (1973). *The interpretation of cultures*. New York, NY: Basic Books.

Gmelch, G. (2003). *Behind the smile: The working lives of Caribbean tourism*. Bloomington, IN: University of Indiana.

Goodwin, H. (2011). *Taking responsibility for tourism*. Oxford, UK: Goodfellow Publishers Ltd.

Goodwin, H., & Francis, J. (2003). Ethical and tourism: Consumer trends in the UK. *Journal of Vacation Marketing*, *9*(3), 271–284. doi:10.1177/135676670300900306

Granzka, P. R. (2014). *Intersectionality: A foundations and frontiers reader*. Boulder, CO: Westview.

Grimwood, B. S. R., Yudina, O., Muldoon, M., & Qiu, J. (2015). Responsibility in tourism: A discursive analysis. *Annals of Tourism Research*, *50*, 22–38. doi:10.1016/j.annals.2014.10.006

Haraway, D. (1990). *Simians, cyborgs, and women: The reinvention of nature*. New York, NY: Routledge.

Haraway, D. (2014). Situated knowledges and the persistence of vision. In P. R. Granzka (Ed.), *Intersectionality: A foundations and frontiers reader* (pp. 41–47). Boulder, CO: Westview.

Kabbani, R. (1986). *Europe's myths of orient*. London: Pandora.

Kapoor, I. (2004). Hyper-self-reflexive development? Spivak on representing the third world "other". *Third World Quarterly*, *4*, 627–647. doi:10.1080/01436590410001678898

La Flamme, A. G. (1979). The impact of tourism: A case from the Bahama Islands. *Annals of Tourism Research*, *6*(2), 137–148.

Latour, B. (1987). *Science in action*. Boston, MA: Harvard University.

Leslie, D. (2012). Responsible Tourism: Concepts, *Theory, and Practice*. Cambridge, MA: CABI.

Malinowski, B. (1922). *Argonauts of the Western Pacific: An account of native enterprise and adventure in the archipelagoes of the Melanesian New Guinea*. Long Grove, IL: Waveland.

Manning, F. E. (1979). Tourism and Bremuda's black clubs: A case of cultural revitalization. In E. Kadt (Ed.), *Tourism—passport to development?* (pp. 147–176). Oxford, UK: Oxford University.

Marcus, G. E. (1995). Ethnography in/of the world system: The emergence of multi-sited ethnography. *Annual Review of Anthropology*, *24*(1), 95–117. doi:10.1146/annurev.an.24.100195.000523

McBride, A., Hebson, G., & Holgate, J. (2015). Intersectionality: Are we taking enough notice in the field of work and employment relations? *Work, Employment & Society*, *29*(2), 331–341. doi:10.1177/0950017014538337

Menéndez, E. L. (1996). *De algunos alcoholismos y algunos saberes: Atención primaria y proceso de alcoholización*. México, D. F.: CIESAS.

Molinié, A. (2013). La invención del new age andino. In R. De la Torre, C/ Gutiérrez Zúñiga, & N. Juárez Huet (Eds.), *Variaciones y apropaciones latinoamericanas del New Age* (pp. 285–308). Mexico City, MX: Casa Chata.

Mooney, S. (2018). Illuminating intersectionality for tourism researchers. *Annals of Tourism Research*, *72*, 175–176. doi:10.1016/j.annals.2018.03.003.

Nader, L. (1972). Up the anthropologist: Perspectives gained from studying up. In D. H. Hymes (Ed.), *Reinventing anthropology* (pp. 1–28). New York, NY: Pantheon Books.

Palmer, C. (1994). Tourism and colonialism: The experience of the Bahamas. *Annals of Tourism Research*, *21*(4), 792–811. doi:10.1016/0160-7383(94)90084-1

Picard, M. (1991, March). Cultural tourism in Bali: National Integration and Provincial Differentiation. In *ASESUK Annual Meeting*, Hull: University of Hull.

Pickel-Chevalier, S., Bendesa, I. K. G., & Putra, I. N. D. (2019). The integrated touristic villages: An Indonesian model of sustainable tourism. *Tourism Geographies Journal*, 1–25. doi:10.1080/14616688.2019.1600006

Rapp, R. (2000). *Testing women*, testing the fetus: The social impact of amniocentesis in America. New York, NY: Routledge.

Said, E. W. (1978). *Orientalism*. London: Penguin.

Sin, H. L. (2010). Who are we responsible to? Locals' tales of volunteer tourism. *Geoforum*, *41*(6), 983–992.

Smith, D., Lommerse, M., & Metcalfe, P. (2014). *Perspectives on Social Sustainability and Interior Architecture Life from the Inside* (2nd ed.). Singapore: Springer Singapore

Spivak, G. C. (1988). Can the subaltern speak? In C. Nelson & L. Grossberg (Eds.), *Marxism and interpretation of culture* (pp. 271–313). Chicago, IL: University of Illinois.

Spivak, G. C. (2003). A Conversation with Gayatri Chakravorty Spivak: Politics and the Imagination." *Signs: Journal of Women in Culture and Society*, *28*(2), 609–624. doi:10.1086/342588

Spivak, G., & Harasym, S. (1990). *The post-colonial critic: Interviews, strategies, dialogues*. New York, NY: Routledge.

Tucker, H., & Hayes, S. (2019). Decentering scholarship through learning with/from each 'other. *Tourism Geographies*, 1–21. doi:10.1080/14616688.2019.1625070

Urry, J. (1990). *The tourist gaze*. London: Sage.

Vega, R. (2017a). Commodifying Indigeneity: How the Humanization of Birth Reinforces Racialized Inequality in Mexico. *Medical Anthropology Quarterly*, *31*(4):499–518. doi:10.1111/maq.12343.

Vega, R. (2017b). I(nter)dentificación racial: racialización de la salud materna a través del Programa Oportunidades y clínicas gubernamentales en México. *Salud Colectiva* (Spanish translation). *13*(3):489–505. doi:10.18294/sc.2017.1114

Vega, R. (2018). *No Alternative: Childbirth, Citizenship, and Indigenous Culture in Mexico*. Austin, TX: University of Texas Press.

Volkman, T. A. (1990). Visions and revisions: Toraja culture and the touristic gaze. *American Ethnologist*, *17*(1), 91–110. doi:10.1525/ae.1990.17.1.02a00060

Weeden, C. (2013). *Responsible tourist behavior*. New York, NY: Routledge.

Wheeler, B. (1991). Tourism's Troubled Times: Responsible Tourism is Not the Answer. *Tourism Management*, *12*(2), 91–96.

Williams, T. R. (2012). Tourism as a neo-colonial phenomenon: Examining the works of Patullo & Mullings. *Caribbean Quilt*, *2*, 191–200.

Wilson, A. (2004). *The intimate economies of Bangkok: Tomboys, Tycoons, and Avon Ladies in the Global City*. Berkeley: University of California Press.

Wong, A. (2015). Caribbean island tourism: Pathway to continued colonial servitude. *Tourisme des îles de la Caraïbe : Le cheminement de l'asservissement colonial 31-32*, (Août- Décembre).

Yan, G., & Almeida Santos, C. (2009). "CHINA, FOREVER": Tourism discourse and self- orientalism. *Annals of Tourism Research*, *36*(2), 295–315. doi:10.1016/j.annals.2009.01.003

Yuval-Davis, N., (2006). Intersectionality and feminist politics. *European Journal of Women's Studies*, *13*(3), 193–209. doi:10.1177/1350506806065752

Zhang, Y., Lee, T. J., & Xiong, Y. (2019). Residents' perception of the influence of luxury chain hotels on ethnic tourism communities. *Journal of Tourism and Cultural Change*, *17*(4), 432–448. doi:10.1080/14766825.2019.1591700

# Exclave accessibility and cross-border travel: The pene-exclave of Ceuta, Spain

Ioulia Poulaki, Andreas Papatheodorou, Alexandros Panagiotopoulos and Sotiroula Liasidou

**ABSTRACT**

Exclaves are characterized by several advantages and disadvantages related to physical, economic, historical, social and cultural attributes; these depend on the owner- and the host-state as well as on relationship between the two. Using the case of Ceuta, a Spanish pene-exclave located in north-western Africa bordering Morocco, air travel mobility and accessibility issues faced by residents are identified and discussed primarily in the context of outbound tourism. The empirical research uses a mixed method approach of data collection based on questionnaires distributed to Ceuta residents and participant observation of the local community. Findings suggest that while enhancing accessibility is generally considered to be positive for travel, the reality is more complex in the case of exclaves and pene-exclaves, where geographical discontinuity and lack of land transport connections with the mainland and hinterland of the owner state may co-exist with geo-political conflicts between the owner and the host countries, i.e. Spain and Morocco respectively in the present context. Based on the concept of tourism for peace and its implications for border tourism and the scaling-up of spatial interaction between neighboring regions belonging to different countries, simpler border-crossing procedures along with innovative destination management policies at local level are proposed to improve exclave accessibility and travel mobility in practice to the benefit of both local residents and inbound tourists.

摘要

飞地依据其物理、经济、历史、社会和文化属性有其优缺点。这些优缺点取决于飞地所属国和东道国以及两者之间的关系。本文以休达(一个位于非洲西北部与摩洛哥接壤的西班牙飞地)为例, 在出境游的背景下确定和讨论了居民面临的航空旅行流动性和可达性问题。实证研究采用以向休达居民发放问卷和对当地社区参与观察为基础的混合数据收集方法。研究结果表明,尽管可达性提升通常被认为对旅行来说是积极方面,就飞地和全飞地（pene-exclaves）来说, 现实情况更为复杂, 那里地理不连续, 与飞地所有国缺乏陆地交通连接大陆和腹地, 可能与飞地所属国和东道国之间的地缘政治冲突共同存在,例如西班牙和摩洛哥目前冲突。基于旅游和平的概念及其对边境旅游、分属不同国家的临近地区空间互动的按比例增加的启示，本文提出在地方层面简化过境手续以及创新目的地管理政策，以改善飞地的可达性和旅游流动性，从实务上惠及入境游客和当地居民。

## Introduction

Regional development, trade and tourism are largely dependent on accessibility often in non-linear ways (Koo & Papatheodorou, 2017; Fujita & Krugman, 2003) characterized by two-way causality (Brugnoli et al., 2018). In the presence of severe discontinuity and isolation because of either physical (e.g. islands) and/or political (e.g. border regions) geography, securing and improving accessibility may prove important for economic prosperity and social welfare. Exclaves and pene-exclaves are a good example of such areas. An exclave is a small piece of a country (i.e. the owner), fully surrounded by the territory of another one, i.e. the host. Moreover, it cannot share a frontier with any other country or be an independent entity; and it will most probably never be self-governed (Gelbman & Timothy, 2011; Vinokurov, 2007; Suzuki & Inohara, 2015). A pene-exclave refers to an outlier slightly connected to its homeland or not fully surrounded by the other country, but accessible by surface transport, only through the territory of the latter (Gelbman & Timothy, 2011; Vinokurov, 2007). This is different from a quasi-exclave which may no longer serve as an exclave (Robinson, 1959). Exclave societies try to balance their existence between the owner state and the host one often facing physical and political barriers. The host country's network and transport infrastructure may prove significant for the exclave's society daily routine facilitation, but political and administrative issues provoked by the occasional conflicts between the two states (owner and host) often lead societies to seek alternatives. In fact, Catudal (1972) discussed the 1971 exchange of West Berlin exclaves with East German enclaves as a unique case of peaceful resolution of such a contested territorial matter.

Tourism may have an important economic and social impact on exclaves (Timothy, 1995, 1996, 2000). In Europe, most exclaves relate to owner and host states both belonging to European Union (e.g. Baarle-Hertog/Nassau a Belgian exclave in the Netherlands) and thus, physical, political and social boundaries are easy to overcome to the benefit of tourism mobility and development. Nevertheless, there are exclaves with EU member as owner state and non-EU member as host such as the Spanish autonomous cities of Ceuta and Melilla surrounded by Morocco in North Africa, where, unresolved political and administrative issues may hinder cross-border travel as well as inbound and outbound tourism (Eckert, 2011; Lagiewski & Revelas, 2004); although, cross-border travel restrictions are evident globally (e.g. North and South Korea – see Prideaux & Kim, 2018; Republic of Cyprus and the north of the island – see Stergiou, 2016), the impact of such issues is exacerbated in exclaves due to their small size and territorial dependency on their host.

Not surprisingly, mobility in tourism has been studied extensively in the literature (Chen & Chang, 2015; Cohen & Cohen, 2015; Coles, Hall & Duval, 2006; Eckert, 2011;

Hall, 2015). Mobility refers first to the ability to move, overcoming physical and political/legal impediments, and second to the direction of moving per se (Gelbman & Timothy, 2019). Leaving political/legal barriers aside, an improvement in accessibility thanks to new transport infrastructure is a necessary but not a sufficient condition to increase tourism mobility, i.e. spatial interaction explicitly depends on travel demand, which may not be easy to generate and/or to sustain in peripheral regions with low levels of perceived tourism potential (Fageda et al., 2018; Hazledine & Collins, 2011; Prokkola, 2007). As infrastructure relates to sunk (i.e. non-recoupable) costs, it may, therefore, prove commercially sensible to improve accessibility of remoter regions by relying initially on air transport, which requires lower levels of infrastructure vis-à-vis surface transport. In addition, the latter is not applicable in islands located too far from the mainland to connect by bridges. Sea transport may offer a much better alternative to air scale-wise, but ships are much slower than airplanes and cannot be used in land-locked peripheral areas. Therefore, prioritizing air transport is strategically important to address accessibility problems of remoter regions and facilitate trade and tourism (Liasidou, 2013; Poulaki & Papatheodorou, 2016; Spasojevic et al., 2018). Indicatively, Abreu et al. (2018) while referring to the Canary Islands, characterize air transport as essential to overcome insularity and assist supply-chain operations, mobility of inhabitants, and the prosperity of the local tourism sector. Intermodal solutions may also prove effective to extend an airport catchment's area in remoter regions located in its vicinity (Poulaki et al., 2017).

The above may be interestingly applied in the context of Ceuta, a pene-exclave and Spanish autonomous city with a population of 82,000 citizens bordering the Moroccan region of Tangier-Tetouan-Al Hoceima, close to the Straits of Gibraltar and half-surrounded by the Mediterranean Sea. Two main reasons justify the study: the first relates to the remoteness of Ceuta given the spatial discontinuity and lack of seamless transport connections between Ceuta and the Spanish mainland (Poulaki & Papatheodorou, 2016). The second concerns its relative tourism potential mainly due to the fruitful blend of the European and Arab worlds. In 2017, Ceuta recorded 75,800 tourist arrivals (i.e. almost equal to its population) with 63% originating from Spain and 37% from other countries (Procesa, 2016). Official hotel accommodation in Ceuta includes 13 units and a total of 782 beds with 2.3 overnights on average (Regional Government of Alicante, 2018).

Accessing Ceuta may prove cumbersome, since traveling internationally requires either a ferry connection and road trip to the closest Spanish airport or a cross-border road trip to the Moroccan airports of Tetouan or Tangier (Figure 1). The latter is a very good alternative from a physical geography point of view but crossing the Hispano-Moroccan borders is subject to lengthy procedures set from both countries. This may have negative repercussions not only for inbound tourism but also for outbound one to the detriment of residents who may feel geographically isolated.

This research aims to identify domestic and international mobility issues in Ceuta building on the following narrative:

- exclaves are a unique feature of socio-political geography often at odds with surroundings areas thus attracting wanderlust and cross-border tourists;

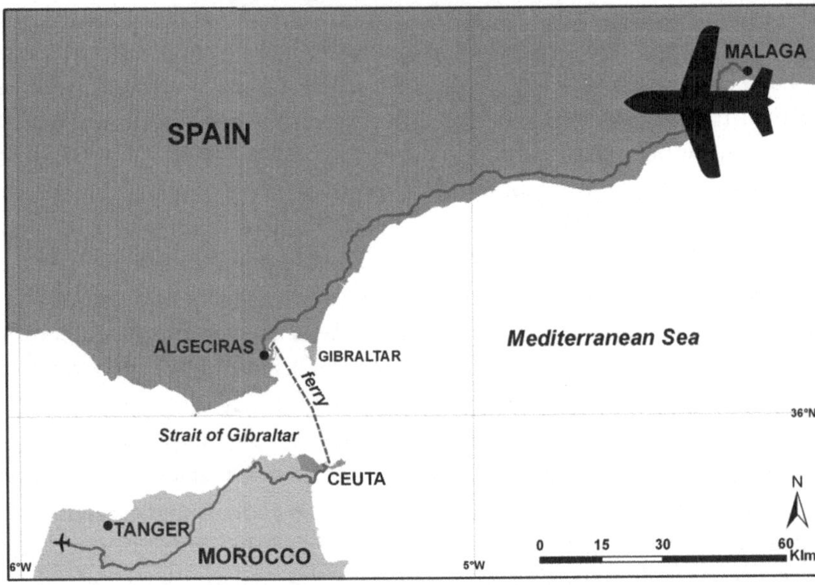


**Figure 1.** Accessing Malaga (AGP) and Tangier (TNG) Airports. Note: The relative size of aircraft in AGP and TNG reflects the volume of 2017 passenger traffic. Source: Authors - map has been produced exclusively for this paper.

- exclave resident social welfare and preservation of cultural identity largely depend on good (air) transport accessibility and seamless connections to the home country but also to the wider world; an effective use of the host country's transport network and infrastructure may, therefore, prove essential for outbound travel;
- such use may be hindered by political tensions between the home and the host country and cross-border impediments; thus, a political détente to improve accessibility is necessary and should be incentivized by highlighting the peace dividend of tourism from an inbound and outbound travel perspective.

In particular, the study seeks to answer the following research questions:

- How do Ceuta's residents travel and how do political issues affect their willingness to do so via Morocco?
- How can the facilitation of border-crossing affect inbound and outbound tourism in Ceuta?

The remainder of the paper first reviews the theoretical nexus among tourism, mobility, air transport and accessibility in general and in the context of exclaves. The case of Ceuta is then discussed; the empirical methodology combines a discrete choice analysis of primary data collected from questionnaires and the results of participatory observation regarding the preferred travel patterns of Ceuta residents. Suitable policy recommendations are then made to alleviate accessibility difficulties faced by Ceuta: a relaxation of Hispano-Moroccan border constraints may set the fundamentals to increase inbound tourism thus rendering financially viable regular shuttle services between Ceuta and Tangier Airport to the benefit of Ceuta residents' outbound travel.

### *Accessibility, air transport, tourism and exclaves*

In addition to the availability of transport modes and related infrastructure, accessibility depends on factors such as frequency of services and fare levels. The latter reflect the cost of service but also the competitive conduct among market participants especially in non-regulated markets (Papatheodorou, 2002). Tourism flows may be enhanced by improved accessibility and therefore tourism development strategies implemented by governments or local authorities' policymakers usually include the construction and/or upgrade of transport infrastructure. Likewise, when tourist demand for a destination increases, then current transport infrastructure should be improved to meet the momentum and sustain destination competitiveness (Crouch, 2011). Crouch and Ritchie (1999) advocate that the success of tourism is intertwined with the performance of sectors such as transportation, contributing to the prosperity and quality of life of the local community too. For remote regions suffering from geographical discontinuity and limited financial resources, transport systems development and thus, improved accessibility, may support lifeline services and lead to new paths of economic growth thanks to tourism mobility. As argued before, air transport may play a critical role in this context; therefore, governments may introduce financial schemes to facilitate air transport services in remoter regions.

In Europe, Public Service Obligations (PSOs) is a documented policy (concerning services, frequencies, scheduling and fares) aiming to provide air connectivity to places that under normal circumstances would have lacked them (Fageda et al., 2018). European PSO routes are characterized by high heterogeneity and discretional criteria among countries in terms of subsidization and market protection. For example, Spanish PSOs coexist with a scheme of passenger-based policies with fare discounts to residents. With Law 66/1997 article 103, routes to Balearic Islands, Ceuta and Melilla joined the scheme. In addition to any minimum service requirements, discounts to residents of Canary Islands, Balearic Islands and several Autonomous Communities amounted to 75% of the full fare in 2017. Fageda et al. (2018, p. 72) underline that 'the resident discount policy in Spain implies an amount of resources higher than the whole European program of PSOs'. Furthermore, non-residents experience higher fares; this may negatively affect inbound tourism flows in remote regions.

Poulaki et al. (2013) show that this may prove the case for the Greek island of Chios located in the Eastern Aegean Sea, whose accessibility may improve by relying more on the airport of Izmir in Turkey that offers a wide network of international connections. In fact, the island is connected to the port of Çeşme on the Turkish coast via a Short Sea Shipping Link, i.e. a ferry service of about thirty minutes. Adnan Menderes International Airport (international code: ADB) of Izmir can then be reached from Çeşme in about 45 minutes by car. On the other hand, to access Athens International Airport (ATH), a Chios islander spends more time and money, since the entire Aegean Sea needs to be crossed by either air or sea. Thus, travel alternatives from Chios to international destinations via ADB seem more attractive. Ceuta resembles Chios as what matters is a political détente between an EU (i.e. Spain, Greece) and a non-EU member state (i.e. Morocco, Turkey) with aspirations to be in good relations with the EU. An improvement of bilateral relations may reduce negativity sentiments and streamline cross-border operations without violating the rules of the Schengen Treaty (both Spain and Greece are members of this Treaty creating a borderless zone within Europe and setting common visa and other procedures vis-à-vis third countries).

Since the liberalization of the air transport market in Europe the number of airports providing commercial operations has risen substantially and passengers have a wider choice of airfares and airports than ever before (Arvanitis & Papatheodorou, 2015). Direct competition among airports has generated more than one air travel alternatives in several cases. Passenger airport choice has been extensively analyzed in the literature (Garrow, 2010; Kouwenhoven, 2008; Lieshout, 2012; Poulaki et al., 2017) with total travel time (including airport accessibility time); total travel cost (including airport accessibility fare); destinations served; and flight frequencies identified as the dominant factors of airport choice. Poulaki and Papatheodorou (2016) underline that an integrated intermodal transport system, with seamless connections of different public transport modes, positively affects an airport's catchment area. Nevertheless, the prioritization of factors determining airport choice depends on geographical conditions and varies among passenger social groups.

Comparative studies of access time and offered services among airports may help illustrate airport market power in a region compared to the rest of the airports located there. The cross-border catchment area of airports located in European regions is a somewhat understudied area (Paliska et al., 2016; Poulaki et al., 2013). Passengers' perception of political borders, linguistic, cultural, societal barriers and different historical/ideological backgrounds (Stoffelen & Vanneste, 2018) may trigger suspicion and mistrust (Altinay et al., 2002; Prideaux & Kim, 2018) thus negatively influencing passenger mobility and willingness to change travel behavior especially with respect to outbound travel. Therefore, focusing on the perception of borders and the subsequent role of tourism for peace is essential to understand and overcome travel inhibitions and complexities. Farmaki (2017, p. 532) indicates an emerging body of cases that present tourism as 'a contributor in reconciliation efforts in post-conflict settings, as an agent to establish mutual understanding and sympathy between divided communities and a confidence-building measure encouraging cooperation'. The tourism and peace nexus in the literature has divided scholars between advocates calling tourism 'The World Peace Industry' and their opponents who ask researchers not to praise unreal virtues (Litvin, 1998). The critical issue between these two scholar groups is whether tourism is a cause or a beneficiary of peace (Farmaki, 2017).

In fact, political boundaries in a region create unique political, economic and social conditions. Special forms of tourism may develop, such as shopping opportunities due to lower prices, or because several activities are permitted on one side of a border but not on the other, due to legal or social norms. Gambling may be such an example when nationals are not allowed to play in their country of residence or when gambling in general is prohibited (Eckert, 2011; Felsenstein & Freeman, 2001; Timothy, 2000). Thus, the neighboring country can act as the place that develops casinos as in the case of China and Macau SAR. In addition, crossing the borders can be a reason for alcohol consumption as in the case of Mexico – USA (Cherpitel et al., 2015) and Saudi Arabia – Bahrain.

Not surprisingly, the nexus among air transport, accessibility and tourism may prove more complex in the case of exclaves. In terms of political integration, exclaves may generate economic or administrative inefficiency mainly because of increased costs of public services provision including transportation and communication. They may also emerge as a weak link from a national defense perspective. On these grounds, the exclave economy has three main options for development: tighter ties with the host state if allowed given extant border constraints; closer relationships with the home state; or strengthening of the local economy, with tourism development playing an important role (Gelbman & Timothy, 2011).

Timothy (1995) regards tourism as the most important industry for several European and North American exclaves. In fact, these are unique from a wanderlust perspective because of the amalgamation of different cultures and community values within a limited space (Gelbman & Timothy, 2011; Hall, 2015). Exclaves are popular as tourist destinations especially to the residents of the host country. However, border permeability is a determinant factor of such a relation. Timothy (1995) identifies two main determinants of tourism mobility in exclaves, i.e. the degree of difficulty for tourists to cross the border; and the degree of similarity in the culture or societies located on both sides of the borders. In any case, exclaves face several problems in developing tourism. Their small size tends to limit their tourism potential in terms of both resources and infrastructure; moreover, they are unable to enact their own tourism development policies since they are not sovereign.

## Accessing the autonomous city of Ceuta

Spain is administratively divided into eighteen autonomous communities with Ceuta being one of them. Nonetheless, remoteness of Ceuta from mainland Spain has inevitably led the government to seek economic resources to boost the city's community welfare. The Spanish government offers higher salaries to public servants who are residents of Balearic Islands, Canary Islands, Ceuta and Melilla. In the latter two cases, salaries are 24.3% higher than those in the capital community of Madrid (Unión Generál de Trabajadores, 2018). Moreover, Ceuta is entitled to receive national and European Union's funds like every other region. With a total budget of 54.7 million euros, of which 43.7 million euros are EU-funded, the Operational Program of Ceuta for 2014–2020 aims to help the city alleviate the challenges it faces as a result of being located in the northern coast of Africa and reduce the difference between the social and economic metrics of the city and the EU average. Around one third of this budget is planned to be invested in transport infrastructure, to decongest daily commercial and tourist traffic and reduce by 50% the traffic jams on the main road that starts at the border with Morocco (European Commission, 2015).

Half surrounded by Moroccan territory, Ceuta can only be accessed by water from the Spanish mainland, i.e. crossing the Straits of Gibraltar by ferry to the port of Algeciras in Andalucía in the southern part of Spain. Moreover, four airports are located within a radius of 200 km (consisting the upper distance limit of an airport's catchment area according to Lieshout, 2012) from Ceuta as shown in Table 1.

The international airport Costa del Sol of Malaga (AGP) is located 171 km from Ceuta and can be accessed by combining a ferry service to the port of Algeciras (which is available all-year round in high frequency as a short sea shipping link) with a

**Table 1.** Airports in the vicinity of Ceuta.

| Airport name | Accessibility from/to Ceuta | | | | Destinations served |
|---|---|---|---|---|---|
| | Distance | Transport Modes | Time | Process | |
| AGP – Costa del Sol Malaga, Spain | 171 km | Ferry & car/ shuttle bus | 3h 32min | Ferry & tolls | >120 (Spain, Morocco & Europe) |
| XRY – Jerez de la Frontera, Spain | 150 km | Ferry & car/ shuttle bus | 3h 18min | Ferry | 11 (Spain & Germany) |
| TNG – Ibn Batuta Tangier, Morocco | 86 km | Car/taxi | 1h 27min | Border crossing & tolls | >30 (Morocco, Spain & Europe) |
| TTU – Sania Ramel Tetouan, Morocco | 38 km | Car/taxi | 42min | Border crossing | 5 (Morocco, the Netherlands & France) |

Sources: www.aena.es, www.onda.ma, www.googlemaps.com, www.rome2rio.com – processed by authors.
Notes: At present there are no scheduled shuttle bus services between Ceuta and neighboring Moroccan cities. Taxi cost is considered as relatively low compared to European standard fares, since a roundtrip from Ceuta to Tangier airport costs 30 euros per person.

trip by car or shuttle bus paying tolls and spending about three hours and thirty minutes. AGP serves more than 120 destinations to Spain, Morocco, Europe, Asia and the Americas. For several years, AGP was also accessed by Ceuta citizens via helicopter flights from the local heliport (code: JCU). This was built by Aeropuertos Espanoles y Navegacion Aerea (AENA), the managing body of Spanish airports, and became operational in 2004 to serve Ceuta citizens being the only heliport to host scheduled domestic flights in Spain connecting two continents. Ceuta citizens were offered a 50% fare discount to travel from JCU to AGP or Algeciras heliport; nonetheless, in July 2014 the helicopter services were suspended due to insufficient demand and high operational costs (Poulaki & Papatheodorou, 2016). Since then, operations start and stop according to the plans made by the companies that undertake the flights. Even after the helicopter flights suspension, international travel from/to Ceuta is mainly produced via AGP. Jerez de la Frontera airport (XRY) in mainland Spain is farther away from AGP and similarly accessible albeit without tolls. The airport offers a very limited network of routes, serving only Spanish metropolitan cities and nine German destinations in seasonal scheduled operations or charter flights.

The other two airports within a 200 km radius from Ceuta are in Morocco being more easily accessible by surface transport albeit with the challenge of border crossing. The international airport Ibn Batuta of Tangier (TNG) is located 86 kms far from Ceuta and can be reached in one hour and twenty minutes-drive by crossing the Hispano-Moroccan borders and paying tolls. Morocco signed an Open Skies agreement with the European Union in 2006 and since then TNG has developed a noteworthy network currently covering over thirty destinations in Morocco, Spain and main European cities. TNG traffic has experienced a spectacular increase of 226% in the past decade mainly due to flights operated by low fare airlines such as Ryanair and EasyJet (Office National Des Aéroports (ONDA), 2018). Finally, Sania Ramel airport in Tetouan (TTU) is located 38 km away from Ceuta and can be accessed in about 40 minutes-drive by crossing the borders. However, its route network is very limited as regular flights are offered to one Moroccan destination only; moreover, there are several seasonal and charter flights to France and the Netherlands.

As shown in Figure 2, ferry traffic between Ceuta and Algeciras is about 2,000,000 passengers per year; traffic peaked in 2015 immediately after the suspension of the

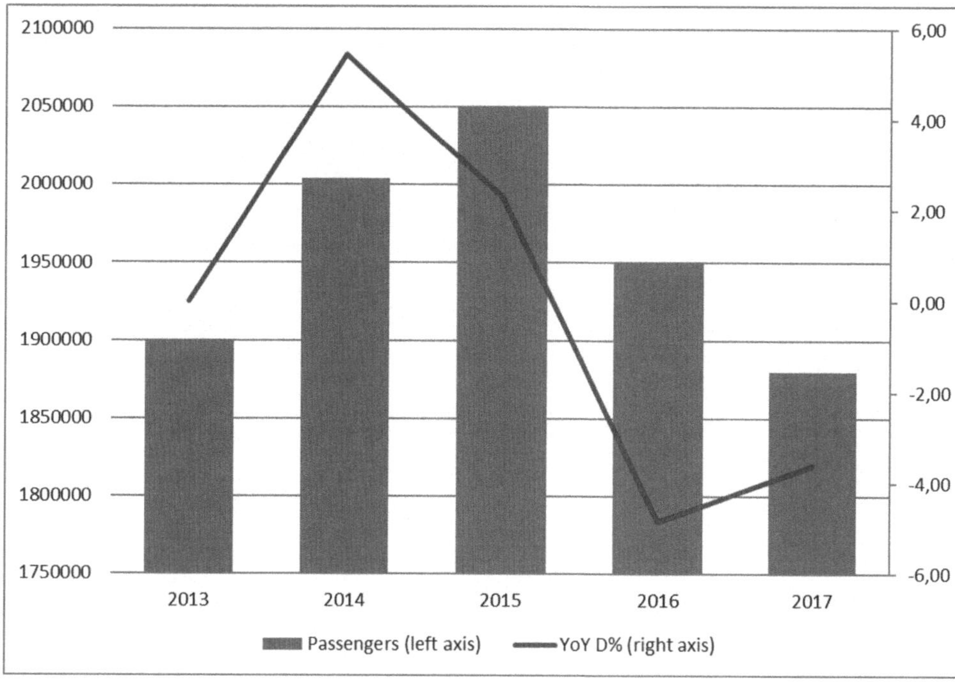


**Figure 2.** Ferry Traffic on the Algeciras-Ceuta Line 2013–2017. Source: Puerto de Algeciras (2018), processed by authors.

regular helicopter services. Therefore, and considering the population of the pene-exclave, it may be concluded that there is significant mobility between Ceuta and the Spanish mainland for commercial, social or leisure purposes including tourism and international travel. On the other hand, the two main airports that serve the wider region (i.e. AGP and TNG) show differences in their profiles in terms of both passenger traffic and network size (Figures 3 and 4). Illustratively, AGP recorded 18,628,876 passengers in 2017, i.e. about 17.4 times more than TNG, which recorded 1,070,247. The traffic ratio between AGP and TNG peaked in 2016, with a value of 19.6 as the annual traffic growth rate of the former was almost twice the one of the latter (i.e. 15.7% vis-à-vis 7.8%) due a strong year for Spanish inbound tourism. Despite these differences, TNG serves most of Europe's main cities, presents significant annual traffic increase and is easier to access from Ceuta.

## Empirical methodology

Studying the outbound tourism behavior of Ceuta's residents to evaluate preferences and perceptions when it comes to airport choice between AGP and TNG is interesting: synergies and conflicts between physical and political geography are expected to play an important role in shaping mobility of residents regarding both domestic and international air flights to and from the exclave city of Ceuta. The study is exploratory in nature (McNabb, 2010) as detailed investigation is required to understand the dynamics of cross-border flows in exclaves. In fact, this study builds on Poulaki and Papatheodorou (2016) who interviewed Ceuta's transport and tourism major

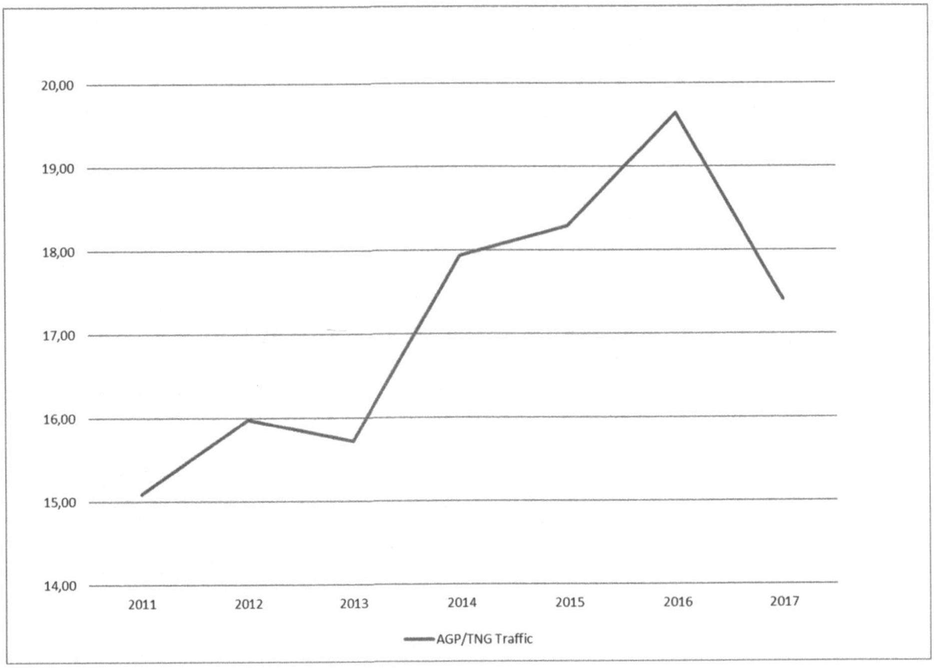


**Figure 3.** AGP/TNG Relative Traffic 2011–2017. Source: AENA (2018), ONDA (2018). Note: AGP stands for Malaga Airport; TNG for Tangier Airport.

stakeholders on their perception of TNG for outbound travel. Their findings suggest that there is a passenger flow from/to Ceuta to/from TNG regarding both domestic Spanish and international destinations and that under certain circumstances the Spanish exclave can be included in TNG's catchment area in favor of tourism development. The present study complements Poulaki and Papatheodorou (2016) by surveying Ceuta's residents on outbound travel behavior and airport choice.

The survey was conducted at the end of July 2017 within one week by one of the authors. Randomly selected participants in social gathering places and the port station were provided details on the research assured of data confidentiality and anonymity. A total of 220 questionnaires were collected while 185 proved valid for statistical analysis. The sample could have been admittedly larger, but many people refused to participate in the (rather lengthy) survey due to lack of time. Despite the random selection of participants, the sample includes representatives from a wide range of socio-demographic groups because the author visited numerous and functionally and/or stylistically very different gathering places for survey purposes; thus, selection bias has been avoided.

Outbound airport choice is tested based on a set of travel behavior variables along with the demographic characteristics of the sample in combination with various alternative travel scenarios from AGP and TNG. The primary data collection involves a mixed-method approach. The first is based on the distribution of questionnaires (mainly with close-ended questions) in Spanish to Ceuta residents and the second refers to participatory observation. Data collected from the questionnaires were analyzed statistically and econometrically using discrete choice modeling, to highlight

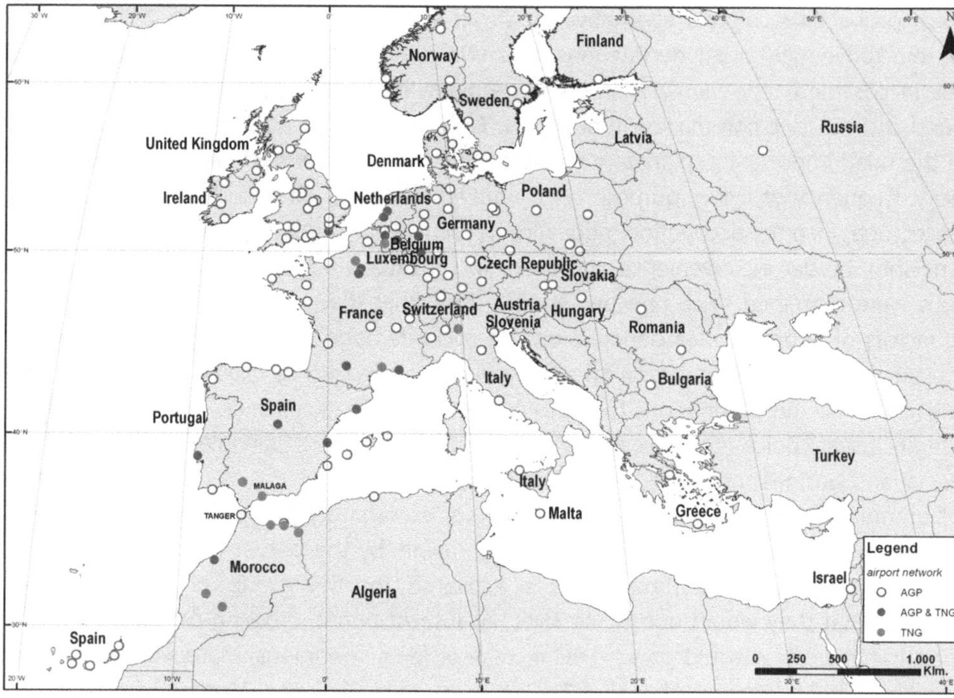


**Figure 4.** Malaga (AGP) and Tangier (TNG) Airport Network.
Source: AENA (2018), ONDA (2018) – map has been produced exclusively for this paper. Note: Regarding Istanbul, Malaga Airport (AGP) is connected to IST (Istanbul Ataturk Airport), while Tangier Airport (TNG) serves SAW (Sabiha Gokcen airport on the eastern side of Bosporus).

utility maximization among travel alternatives per social group of participants depending on their demographic characteristics.

Respondents to the questionnaire were asked to choose the air travel/airport option that maximized their utility. Door-to-door roundtrip scenarios were built using the 10 July 2017 as the hypothetical booking date and the last week of September (i.e. 20–27/9/2017) as the dates of the hypothetical trip. These dates were selected for specific reasons: TNG and AGP are two airports with high seasonality due to a summer peak period, thus, the end of season, i.e. late September, can combine both attributes, i.e. availability of connections and the existence of relatively affordable leisure fares. Several online booking engines were used to compare possible intermodal combinations thus providing the participants with the most effective travel alternatives (cost- and time-wise) from Ceuta to ten main destinations within Europe and the Americas (all served by both AGP and TNG), avoiding night stops in airports. As shown in Appendix 1, the chosen set included Madrid and Barcelona as the two main business centers in Spain; Amsterdam, Brussels, Frankfurt, Lisbon, London and Paris based on the preference set of Spanish outbound travelers according to the Hotel Price Index (2018); as well as New York and Buenos Aires, which in addition to acting as important final destinations for Spanish travelers to the Americas (for business, leisure and visiting-friends-and-relatives (VFR) purposes), they are also important connecting points to other places in the continent.

A discrete choice analysis was undertaken (Garrow, 2010) using Biogeme software version 1.8 based on a multinomial logit (MNL) model. Dependent variables include the choice probabilities of two alternative air travel scenarios for a roundtrip from Ceuta to the ten above-mentioned destinations. The two alternatives relate to the optimal scenarios in cost and time via AGP and TNG respectively. Based on the theory behind outbound tourism decision-making (Boukhrouk & Ed-Dali, 2018), travel cost and travel time are the most important explanatory variables to consider; others relate to destination specificity, frequency of flying, purpose of trip, travel company, direct/indirect travel booking, gender, age, income, occupation, education level, marital status, permanent residence, car ownership, border inconvenience and aversion to travel via Morocco. Gender and marital status were examined since previous studies show that the prioritization of the determining factors of airport choice depends on the socio-demographic passenger characteristics (Poulaki et al., 2017). Destination specificity was modeled using related dummy variables for each of the ten destinations to capture any unique characteristics. Frequency and purpose are usual choice variables entering the model as dummy variables (i.e. low/high frequency and business/non-work-related travel respectively).

Dummy variables were also used for all other explanatory variables. The discomfort of getting from Ceuta to Tangier Airport (caused mainly by the border crossing) was quantified by introducing a dummy variable activated for the group of individuals who answered that they would not fly via TNG because of border crossing or lack of transportation but that they would prefer TNG in case of ideal conditions. Moreover, people who seem to be biased against choosing TNG in every case (i.e. being Morocco-averse in general) were associated with a separate dummy variable. Before any estimation, missing values were purged from the dataset; all response entries were considered unique, assuming identical error distribution across observations and not across individuals.

Participant observation was also used to test the validity of the statistical results from a qualitative perspective. Participant observation provides a better insight on descriptive research questions, to build theory or to generate or test hypotheses (Kawulich, 2005). Also, it provides opportunities for viewing or participating in unscheduled events (Kawulich, 2005); it improves the quality of data collection and interpretation; and facilitates the development of new research questions or hypotheses (Kawulich, 2005). The rationale for having an alternative method of data collection was because of the authors' lack of exposure to the daily life, culture and attitudes of Ceutan residents. Thus, to properly evaluate the data collected from the questionnaire and avoid unrealistic/non-sensical conclusions, one of the authors participated in informal, unstructured discussions with the residents in the border area with Morocco; in places of socializing and at Ceuta's port passenger terminal regarding their outbound travel behavior. Author's fluency in Spanish contributed into easing discussions among the participants and the researcher.

## *Analysis and results*

### *Preliminary statistical analysis*

The sample was balanced on gender and marital status terms. 66.9% of the respondents were between 25 and 50 years old and 58.7% were earning monthly between €1,000 and €3,000. 81.1% of respondents had at least college education. These are in line with Garín-Muñoz and Moral-Rincón (2012), who found that the average Spanish

**Table 2.** Airport choice and main reason of selection.

| | Scenario 1 | | | | Scenario 2 | | |
|---|---|---|---|---|---|---|---|
| Destination | Optimal scenario in terms of time- and cost-effectiveness via AGP | | | | Optimal scenario in terms of cost-effectiveness via TNG | | |
| New York | | | 87.1% | | | 12.9% | |
| London | | | 84.7% | | | 15.3% | |
| Buenos Aires | | | 81.8% | | | 18.2% | |
| Amsterdam | | | 62.8% | | | 37.2% | |
| Madrid | | | 62.7% | | | 37.3% | |
| Barcelona | | | 45.9% | | | 54.1% | |
| Frankfurt | | | 42.2% | | | 57.8% | |
| Lisbon | | | 39.7% | | | 60.3% | |
| Paris | | | 39.4% | | | 60.6% | |
| Brussels | | | 37.0% | | | 63.0% | |
| Main reason of choosing travel scenario | Travel cost 37.1% | Travel time 4.6% | Cost & time efficiency 21.8% | Inconvenient border crossing 16.6% | AGP access long lasting 2% | Don't want to travel via Morocco 12.6% | Other 5.3% |

Source: Authors.
Note: AGP stands for Malaga Airport; TNG for Tangier Airport.

outbound tourist is an employed person aged between 25 and 44 years of at least secondary education and a frequent internet user. Forty-six percentage of the sample's participants take at least one roundtrip by air per year, while 19% take five such roundtrips mainly for leisure and VFR. Only 38.2% had previously flown from JCU to AGP and 53.1% of those who have not flown by helicopter found its fares very expensive. Nonetheless, after the suspension of JCU-AGP services, AGP use by participants increased with 54.4% of them having flown to Barcelona (BCN) and Madrid (MAD) as final destinations. They accessed AGP by combining a ferry boat with either private/rented car (62.2%) or shuttle bus (23.2%), while only 14.6% of the participants declared that they used other transport mode without specifying it. The latter are mainly young and well-educated people, employed by NGOs that contribute to immigration management in the Ceuta region and live permanently in the southern part of Spain.

With regards to the previous TNG utilization by Ceuta residents, 46.9% of them have flown to MAD and 15.6% to BCN as their final destination. 26.6% chose TNG to fly internationally with a direct flight, given the high frequencies and the wide network of destinations offered by the airport. Accessibility to TNG is mainly facilitated by car (61.5%), despite the long car queues in the border crossing. Additionally, 26.2% of the participants have accessed TNG by taxi or 'grand taxi', while few of them had been transferred by private shuttle bus.

Results with regards to travel scenarios alternatives are presented in Table 2, illustrating the outbound travel pattern followed by residents of Ceuta when choosing an airport, ranked in descending significance with respect to AGP. On average 58.3% of respondents chose Scenario 1 (via AGP) and 41.7% Scenario 2 (via TNG). Irrespectively of their chosen scenario, 37.1% on average responded that flying in low budget is an important determinant for their choice with only 4.6% emphasizing predominantly time-saving; a combination of cost and time efficiency was highlighted as the primary reason among 21.8% of participants. A 16.6% indicated that inconvenient border crossing procedures between Morocco and Spain was the primary reason for selecting Scenario 1 although the importance of this inconvenience was highlighted by most respondents. Additionally, 12.6% provided responses in favor of Scenario 1 mainly related to the political conflict between Spain and Morocco and 'ethical' (apparently national-centric) reasons.[1] These may also relate to an ongoing local migration issue on top of the 2015–2016 general migration/refugee crisis in Europe and its implications for tourism (Pappas & Papatheodorou, 2017): in fact, the six-meter high, 8.4 km long border-fence separating Morocco from Ceuta reinforces the image of a 'fortress-Europe' (Castan Pinos, 2009); at the same time, attempted and/or successful border breaches receive wide publicity (BBC, 2017) but also concern at a local level as was the case in 2005 but also in December 2016 and February 2017, i.e. just a few months before the on-site study took place.

In any case, 28% of the participants' identified TNG as the nearest main international airport irrespectively of the chosen travel scenario. Additionally, 20% supported that TNG may not be preferred by many Ceutans because of the lack of shuttle bus with regular services between Ceuta and the airport – lack of secured car parking facilities at TNG was also mentioned as a factor raising concerns. Interestingly, 71% of respondents would have clearly opted for TNG had ideal conditions existed, i.e. almost seamless border crossing and the existence of a shuttle bus connecting Ceuta city center with TNG. As also discussed later in the paper, this is an important finding suggesting that the improvement of accessibility of the pene-exclave from its host country could have very positive repercussions not only for outbound but also for inbound tourism in Ceuta.

Chi-square analysis/Monte Carlo test results reveal that the majority of those who have already traveled via TNG in the past identified TNG as the nearest airport to Ceuta. Moreover, respondents employed in the public sector are less willing to use TNG, possibly because of their national-centric sentiments. Respondents who would not travel via TNG chose the travel scenario via AGP because either they do not want to travel via Morocco or for reasons such as the absence of secure car parking facilities. Place of residence is a statistically significant determinant factor at 10% level (though not at 5%) since those who identified themselves as permanent residents of the Southern part of mainland Spain, chose not to use TNG.

### *Participant observation findings*

Most respondents mentioned the inconvenient border crossing procedure and lack of secure parking facilities as preventing factors when it comes to an airport's selection. Locals shared previous travel experiences in a vivid manner referring to cases where they had to wait up to twelve (12) hours in the car, queuing for permission to cross the border, making border-crossing unbearable. This is also supported by Hasan-Kerr (2018) who refers to additional security controls in vehicles and limited border officers' availability. Moreover, Rijk (2018) points that non-EU citizens can cross the border into Ceuta only if they can present a valid Schengen visa, (except for people residing in the Tétouan region). This inevitably causes very inconvenient congestion in the border control offices. Moroccan passport-holders can cross the border relatively easily if they have documentation proving that they work in Ceuta. Such jobs concern mainly

**Table 3.** Results of the initial multinomial logit model ($n = 1{,}212$ observations).

| Name | Coefficient Value | t-test | *p*-value |
|---|---|---|---|
| $A_{AGP}$ | 0.996 | 3.56 | 0.00 |
| $A_{TNG}$ | 0.000 | n/a | n/a |
| $B\_COST_{AGP}$ | −0.007 | −7.41 | 0.00 |
| $B\_COST_{TNG}$ | −0.007 | −7.37 | 0.00 |
| $B\_TIME_{AGP}$ | −0.188 | −2.71 | 0.01 |
| $B\_TIME_{TNG}$ | −0.183 | −2.98 | 0.00 |

Source: Authors.

Note: Coefficient values have been rounded to three decimal figures. AGP stands for Malaga Airport; TNG for Tangier Airport.

housekeepers, babysitters and workers. It is worth mentioning that Spanish employers are responsible to keep such documents valid, otherwise borders cannot be crossed easily making workers' daily lives difficult. Furthermore, some Ceutans highlighted the negativity (if not hostility) of certain people on the other side of the border as a major point of concern: for this reason, they prefer to travel by sea to the Spanish mainland, crossing the Gibraltar straits, carrying out commercial activities and traveling further from there. Negativity may relate to a different *Weltanschauung* not to mention the long-standing conflict between Spain and Morocco over the status of Ceuta.

### *Results of discrete choice analysis*

In a first, simple model only the travel time and travel cost variables (i.e. the generalized travel cost) are investigated. Each variable is assigned a coefficient to be tested separately and determine whether there is a significant difference in the respondents' perception of the two travel scenarios. Quandt (1970) argues that travelers tend to make transport choices based on the concept of generalized transport cost, which jointly considers the value of time and travel expenses. Thus, the model specification is:

$$S_{AGP} = A_{AGP} + B_{TIME_{AGP}} * TT_{AGP} + B_{COST_{AGP}} * TC_{AGP}$$

and

$$S_{TNG} = A_{TNG} + B_{TIME_{TNG}} * TT_{TNG} + B_{COST_{TNG}} * TC_{TNG}$$

where: $S_i$ is the choice probability of scenario $i$ (AGP or TNG); $A_i$ is a scenario-specific constant; $B\_TIME_i$ is the travel time coefficient of $i$; $TT_i$ is the travel time variable (measured in hours) of $i$; $B\_COST_i$ is the travel cost coefficient of $i$; and $TC_i$ is the travel cost variable (measured in euros) of $i$. The scenario-specific constant is fixed to zero for the second alternative to avoid multicollinearity. This constant effectively captures any preference factor not covered already by the existing variables and thus expresses the 'default' preference of the population toward either choice scenario.

Results are reported in Table 3. The estimated values for the cost and time coefficients are negative, thus demonstrating an expected micro-economic behavior where longer travel time and higher monetary costs negatively affect choice attractiveness.

Moreover, there seems to be a rather insignificant difference of the coefficients between the two scenarios as evidenced by the relevant cross t-tests, i.e. differences in the coefficients of generalized cost between AGP and TNG is close to zero. This suggests that the population perceives equally the cost of time and price for either scenario. For this reason, it was decided to merge the generalized travel cost variables across both scenarios (i.e. $B\_TIME_{AGP} = B\_TIME_{TNG} = B\_TIME$); similarly, for cost coefficient (i.e. $B\_COST_{AGP} = B\_COST_{TNG} = B\_COST$). The remaining explanatory variables were introduced using a general-to-specific approach (Campos et al., 2005) keeping in the final model specification only those statistically significant variables that contributed to the explanatory power of the model and were thus proven relevant to the decision-making process of the given population for the choice problem. In total 1212 observations were included in the dataset, yielding a $R^2$ value of 0.323. Some variables, such as Frequency and Location were eventually omitted because they did not fully satisfy the t-test at 5% level of significance or because their impact was captured by another variable. Having the above in mind, the final MNL model specification is:

$$S_{AGP} = A_{AGP} + B_{TIME} * TT_{AGP} + B_{COST} * TC_{AGP} + B_{TCINC} * TCINC_{AGP} + B_{DESTBRU} * DESTBRU + B_{DESTLON} * DESTLON + B_{DESTBCN} * DESTBCN + B_{DS} * DS$$

and

$$S_{TNG} = A_{TNG} + B_{TIME} * TT_{TNG} + B_{COST} * TC_{TNG} + B_{TCINC} * TCINC_{TNG} + B_{BORDER} * BORDER + B_{NOMAROC} * NOMAROC + B_{CAR} * CAR + B_{AGE2} * AGE2$$

where $S_i$ is the choice probability of scenario $i$ (AGP or TNG); $A_i$ is the scenario-specific constant; the *B*s refer to common coefficients in each case, e.g. *B_TIME* is the common travel time coefficient; $TT_i$ is the travel time variable (measured in hours) of scenario $i$; $TC_i$ is the travel cost variable (measured in euros) of $i$; $TCINC_i$ is the interactive (cost)-*(high monthly income, i.e. > €2,000) variable of $i$; *DESTBRU/DESTLON/DESTBCN* is the dummy variable for destination Brussels, London and Barcelona respectively; *DS* is a dummy variable activated for non-local civil servants who earn a Double Salary; *BORDER* is the dummy variable activated for the border-crossing averse group of people; *NOMAROC* is the dummy variable activated for the Morocco-averse group of people; *CAR* is the dummy variable activated in the case of car ownership; and *AGE2* is the dummy variable activated in the presence of age group 2 people (i.e. 25–35 years old).

Results reported in Table 4 are in accordance with utility maximization principles considering generalized transport cost. Age group 25–35 years (people who are usually well-educated, independent-minded but not necessarily very affluent) seems to favor TNG over AGP. The other age cohorts do not exhibit any statistically significant preference, except for age group 65+ which seems to favor AGP, but in the end was excluded from the specification due to cross correlation issues with other variables. The impact of the border-crossing is evidently negative when choosing TNG, but car ownership is positively related to TNG, i.e. people who own a car consider TNG more as an option, because it allows them to get there more easily, despite the lack of secured car parking facilities at TNG and the otherwise significant discomfort that many people experience when crossing the border. From the destination dummies, only three were kept, since the rest failed the t-test against zero value. The results strongly suggest that travelers to Brussels and Barcelona prefer AGP over TNG. This bias can be explained based on habit (i.e. people have used AGP to fly to BRU or BCN

**Table 4.** Results of the final multinomial logit model ($n = 1,212$ observations).

| Factors that influence the choice | Choice against which factor was tested | Qualitative impact | Coefficient Value (quantified impact) | Variable Unit | t-test | $p$-value |
|---|---|---|---|---|---|---|
| Default Relative Preference | AGP | positive | 0.925 | 1 | 3.42 | 0.00 |
| Impact of Travel Time | AGP,TNG | negative | -0.191 | Hours | −3.12 | 0.00 |
| Impact of Travel Cost | AGP,TNG | negative | -0.010 | EUR | −10.68 | 0.00 |
| Combined Impact of Travel Cost for Higher Incomes (>2000 EUR) | AGP,TNG | positive | 0.005 | EUR* dummy (0/1) | 3.03 | 0.00 |
| Impact of Specific Destination Brussels | AGP | negative | -0.676 | dummy (0/1) | −2.65 | 0.01 |
| Impact of Specific Destination London | AGP | positive | 0.649 | dummy (0/1) | 2.27 | 0.02 |
| Impact of Specific Destination Barcelona | AGP | negative | -0.786 | dummy (0/1) | −3.36 | 0.00 |
| Impact of High Income (non-local Civil Servants earning Double Salary) | AGP | positive | 0.794 | dummy (0/1) | 2.91 | 0.00 |
| Impact of Border Crossing | TNG | negative | -0.672 | dummy (0/1) | −4.36 | 0.00 |
| Impact of Aversion to Morocco | TNG | negative | -3.740 | dummy (0/1) | −8.95 | 0.00 |
| Impact of Car Ownership | TNG | positive | 0.595 | dummy (0/1) | 3.47 | 0.00 |
| Impact of belonging to the Age group 25-35 | TNG | positive | 0.816 | dummy (0/1) | 5.24 | 0.00 |

Source: Authors.
Note: AGP stands for Malaga Airport; TNG for Tangier Airport.

in the past; therefore, they prefer to stick to their existing pattern). On the contrary, London is perceived to be better accessed via TNG, indicating again either a bias rooted in past choices, or perhaps better perceived connectivity overall.

Those who are overall negative to Morocco demonstrate a strong and clear bias against TNG as expected. What is noteworthy, though, is the size of the impact on their choice. The estimated value of the related coefficient surpasses by far any other dummy variable coefficient, implying that the social group in question is inelastic in its preferences; in other words, they seem to be willing to accept higher costs and longer travel times to avoid flying out of Morocco. High income individuals per se seem to prefer AGP over TNG. It must be noted though that this specific estimate should not be regarded as reliable in a global sense perhaps, because a large part of the high-income individuals in the dataset includes non-local civil servants. The seemingly high heterogeneity of the group of people captured by this dummy, possibly distorts the pure income effect. The impact of travel cost should be examined together with the interactive variable TC_INC, to account for the heterogeneous effect of high income. Indeed, B_TCINC demonstrates a positive sign, as opposed to the negative of B_COST, i.e. travel cost plays a relatively smaller role in the decision-making of higher income individuals . Intuitively, this may be explained more clearly when calculating the Willingness-To-Pay (WTP), or Value-of-Time. This is given by:

$$WTP = \frac{\frac{dU}{dTT}}{\frac{dU}{dTC}} = \frac{B_{TIME}}{B_{COST} + B_{TCINC} * INC} = \frac{-0.191}{-0.0102 + 0.00533 * \text{INC}}$$

where INC = 1 in the presence of high monthly income (over €2,000) and is 0 otherwise. Thus, for individuals with monthly income below €2,000 the WTP is:

$$WTP_{low} = \frac{-0.191}{-0.0102 + 0.00533 * 0} = €18.73 \textit{ per hour}$$

For individuals with monthly income exceeding €2,000, the WTP is:

$$WTP_{high} = \frac{-0.191}{-0.0102 + 0.00533*1} = €39.22 \ per \ hour$$

In other words, high income earners have a willingness-to-pay which is double as high compared to low-income earners.

The above results may be used to simulate a hypothetical scenario of improving the connection between Ceuta and TNG, by adding a shuttle service and easing the border crossing. First, the model is simulated for the base case to test whether it accurately reproduces reality. Among the 1,212 observations of the stated preference survey 680 (or 56.1%) favored the AGP scenario and 532 (or 43.9%) favored the TNG one. Based on the estimates, the model predicts a 672 – 540 split (55.4% vs 44.6%) in favor of AGP which is fairly accurate. Nonetheless, to further improve precision the alternative specific constant is recalibrated at a new estimated value of 0.954 instead of 0.925. With this new value the MNL model reproduces 'perfectly' the observed market shares. Subsequently, the value of the dummy *BORDER* is set at zero for all observations to simulate the existence of seamless travel in all cases. The new utility scores are thereby calculated, and the choice probability is derived by applying the MNL formulation:

$$Pr_{AGP} = \frac{e^{U_{AGP}}}{e^{U_{AGP}} + e^{U_{TNG}}}$$

to every respondent. The preferred choice is assumed to be the one with the highest probability. Simulation results suggest that improving the connection between Ceuta and TNG, all else held constant, may lead to a significant diversion of outbound flows in favor of TNG that ends up with a market share of about 52% as opposed to 48% by AGP.

## Discussion and policy recommendations

Transport alternatives provided by the travel scenarios via TNG address issues of geographical discontinuity and indicate the importance of physical versus political distance vis-à-vis tourism development. As previously discussed in the context of Table 2, when considering the actual choices of the respondents, travel scenarios via AGP scored 58.3%. This percentage includes a 16.6% that chose AGP due to inconvenient border crossing and a 12.6% due to 'ethical' (national-centric) reasons. Consequently, the 'home airport' factor proves important as airport choice is not always dependent on individual characteristics but on perceptions of borders and country of origin (Paliska et al., 2016); this is also highly related to the assumptions of Simandan (2016) and Sofield (2006) about the social, ethnical and cultural differences that provoke negativity to the distant 'other'. After all, 'I do not want to travel via Morocco' received a respectful percentage as a justification of airport choice/preference in the study reflecting a negativity toward the other side of the border.

Reasonably, the political conflict between Spain and Morocco over Ceuta's territoriality and the on-going immigration problem may have affected this preference along with social and cultural differences that exist between Spanish and Moroccan people. Under different circumstances, TNG could have gained an additional $16.6 + 12.6 = 29.2\%$ to reach a final share of 70.9% in passenger airport choice. Thus, it can be said that political geography reigns over physical one in the case of Ceuta impeding the enhancement of tourism flows to/from the destinations on either side of the border. In fact, the results of the calibration exercise show that physical geography may take 'revenge' over political geography if the artificial constraints set by the latter (i.e. borders between countries) are somewhat relaxed. Paliska et al. (2016) reached a similar conclusion in the case of the upper Adriatic multi-airport region as countries previously in tension (i.e. Italy, Slovenia and Croatia) are all now part of the European Union. The case of Ceuta sounds simpler from a technical viewpoint as e.g. the establishment of a shuttle service bus between TNG and Ceuta is rather easy and does not require an intermodal intervention (i.e. combination of water and surface transport) as in the case of Chios, Greece. This accessibility improvement may prove to the benefit not only of outbound travelers from Ceuta but also to inbound tourists interested in visiting the Spanish pene-enclave. The very development of cross-border tourism may set the economic fundamentals and generate the necessary scale to render the shuttle bus service financially viable and intensify pressure on politicians on both sides of the border to further relax border-crossing procedures.

To overcome such cross-border problems and by adopting the viewpoint of tourism-for-peace advocators, a potential cooperation is proposed in a common tourism agenda between Spain and Morocco in the border region of Ceuta and Tangier – Tétouan-Al Hoceima. Such an initiative may lead local communities into a different perception of contacting each other and thus, to increased accessibility and tourism development. It can lead to a win-win situation for both parties and welfare since the TNG use by locals and tourists may increase mobility from and to the Spanish pene-exclave.

## Conclusions and the way forward

Border areas have narratives as a storyline showcasing an amalgam of mixed cultures which become an important tourism attraction. Nonetheless, cross-border regions are at a disadvantage because of their peripheral location, usually away from business centers in the home country, and the consequential neglect compared to more central places. Exclaves may face further unique problems because of their small physical size and inevitably limited resources and ability to expand their infrastructure and provision of public services.

This paper investigated accessibility issues in the Spanish pene-exclave of Ceuta located close to the Moroccan border in Northern Africa. Emphasis was given on outbound traveling by Ceutan residents and the determinant factors behind the choice of alternative airports located in the Spanish and the Moroccan territory, namely AGP and TNG. Based on the findings, an easier border-crossing may change passenger airport choice share in favor of TNG, as in terms of physical distance it is by far closer to Ceuta and literature suggests that airport accessibility is a significant factor to affect the passenger's choice. Undoubtedly, accessibility may improve in the case of simpler border procedures, since the time spent to reach a transport infrastructure may heavily affect the generalized transport cost of potential passengers. Moreover, boosting border tourism may benefit all involved stakeholders. By capitalizing on the existing airport transport infrastructure, i.e. the international airport Ibn Batuta of Tangier, frequent shuttle connections may enhance trade and tourism mobility between the two sides of the border and local authorities may seize the opportunity to cooperate and plan strategically to engage tourism enterprises and local shops in an (admittedly ambitious) local inter-state public-private partnership. In fact, a 'Two Countries – One Experience' campaign may promote the geographical advantage of crossing the borders between Ceuta and Morocco to combine two different tourism products in a single holiday trip; a local inter-state Destination Development Management and Marketing Organization could oversee the process.

The study contributes to the tourism geographies literature by highlighting the implications of political geography for accessibility and outbound travel behavior of residents in peripheral areas. Despite their crucial importance, such issues remain relatively underexplored in the context of exclaves. Hence the very choice of Ceuta is duly justified, and the empirical results are of interest to other regions with similar characteristics. Future research may complement existing findings by considering the perceptions and experiences of inbound tourists and/or other involved stakeholders in the exclave. This is important as the partnership initiatives suggested above rely heavily on a political détente in the Hispano-Moroccan relations and a more efficient management of the immigration problem; significant resistance to change may also emerge from stakeholders who benefit from the current regime. The major issue, therefore, would be to qualitatively test the persuasive power of the narrative that prosperity in Ceuta (as in other exclaves and peripheral regions) should be fostered via extrovert exposure instead of introvert protectionism and regional subsidies, so that public resources can be reallocated to the benefit of other important needs in Spain (or in any other country) and the region itself.

## Note

1. The conflict between Spain and Morocco is political-territorial in nature and is still having an influence on the border (Rijk, 2018). More specifically, Morocco has challenged the Spanish dominance in Ceuta as a relic of Spanish colonialism and argues that Ceuta should belong to the Moroccan region of Tangier. On the contrary, for Spain, Ceuta is a territory upon conquest before 1668 when Portugal accepted Ceuta as Spanish territory in the Treaty of Lisbon (Poulaki and Papatheodorou, 2016).

## Acknowledgments

The authors are grateful to Mr. George Tataris, Environmental Cartographer, Civil and Transport Engineer in the Laboratory of Cartography & Geoinformatics of the Department of Geography, University of the Aegean, Lesvos, Greece, for the production of maps in Figures 1 and 4.

## Disclosure statement

No potential conflict of interest was reported by the authors.

## ORCID

Ioulia Poulaki http://orcid.org/0000-0001-6861-7136
Andreas Papatheodorou http://orcid.org/0000-0001-9484-7517
Sotiroula Liasidou https://orcid.org/0000-0002-9080-2132

## References

Abreu, J., Fageda, X., & Jimenez, J. L. (2018). An empirical evaluation of changes in Public Service Obligation in Spain. *Journal of Air Transport Management*, *67*, 1–10. https://doi.org/10.1016/j.jairtraman.2017.11.001

Aeropuertos Españoles y Navegación Aerea (AENA). (2018). *Estadisticas*. Retrieved September 9, 2018, from www.aena.es (in Spanish)

Altinay, L., Altinay, M., & Bicak, H. (2002). Political scenarios: The future of the North Cyprus tourism industry. *International Journal of Contemporary Hospitality Management*, *14*(4), 176–182. https://doi.org/10.1108/09596110210427021

Arvanitis, P., & Papatheodorou, A. (2015). Greek airports in transition: From public ownership to PPP concessions. *Journal of Airport Management*, *9*(3), 284–295.

Autoridad Portuaria de Ceuta (Port of Ceuta). (2018). *Estadisticas*. Retrieved September 9, 2018, from http://www.puertodeceuta.com/ (in Spanish)

BBC. (2017). *Hundreds of migrants storm fence to reach Spanish enclave of Ceuta*. Retrieved April 20, 2020, from https://www.bbc.com/news/world-europe-39002111

Boukhrouk, M., & Ed-Dali, R. (2018). Outbound tourism in developed countries: Analysis of the trends, behavior and the transformation of the Moroccan demand for international travels. *International Journal of Business, Human and Social Sciences*, *12*(7), 866–873.

Brugnoli, B., Dal Bianco, A., Martini, G., & Scotti, D. (2018). The impact of air transportation on trade flows: A natural experiment on causality applied to Italy. *Transportation Research Part A: Policy and Practice*, *112*, 95–107. https://doi.org/10.1016/j.tra.2018.01.010

Campos, J., Ericsson, N. R., and D. F. Hendry, eds. (2005). *General-to-specific modeling*. Edward Elgar Publishing.

Castan Pinos, J. (2009). *Building Fortress Europe? Schengen and the cases of Ceuta and Melilla*. Centre for International Border Research (CIBR), WP18/2009, 1–29.

Catudal, H. M. (1972). The Berlin agreement: An exchange of exclaves and enclaves. *Geoforum, 3*(3), 78–80. https://doi.org/10.1016/0016-7185(72)90089-9

Chen, J., & Chang, T. C. (2015). Mobilising tourism research in emerging world regions: Contributions and advances. *Current Issues in Tourism, 18*(1), 57–61. https://doi.org/10.1080/13683500.2014.932337

Cherpitel, C., Ye, Y., Zemore, S., Bond, J., & Borges, G. (2015). The effect of cross-border mobility on alcohol and drug use among Mexican-American residents living at the U.S.–Mexico border. *Addictive Behaviors, 50*, 28–33. https://doi.org/10.1016/j.addbeh.2015.06.008

Cohen, S., & Cohen, E. (2015). A mobilities approach to tourism from emerging world regions. *Current Issues in Tourism, 18*(1), 11–43. https://doi.org/10.1080/13683500.2014.898617

Coles, T., Hall, C. M., & Duval, D. (2006). Tourism and post-disciplinary enquiry. *Current Issues in Tourism, 9*(4-5), 293–319. https://doi.org/10.2167/cit327.0

Crouch, G. I. (2011). Destination competitiveness: An analysis of determinant attributes. *Journal of Travel Research, 50*(1), 27–45. https://doi.org/10.1177/0047287510362776

Crouch, G. I., & Ritchie, J. R. B. (1999). Tourism, competitiveness, and societal prosperity. *Journal of Business Research, 44*(3), 137–152. https://doi.org/10.1016/S0148-2963(97)00196-3

Eckert, A. M. (2011). Greetings from the Zonal Border tourism to the Iron Curtain in West Germany. *Studies in Contemporary History, 8*, 9–36.

European Commission. (2015). *€612 million of EU regional funds allocated to Valencia and Ceuta*. Retrieved December 30, 2018, from http://ec.europa.eu/regional_policy/en/newsroom/news/2015/07/eur612-million-of-eu-regional-funds-allocated-to-valencia-and-ceuta

Fageda, X., Suárez-Alemán, A., Serebrisky, T., & Fioravanti, A. (2018). Air connectivity in remote regions: A comprehensive review of existing transport policies worldwide. *Journal of Air Transport Management, 66*, 65–75. https://doi.org/10.1016/j.jairtraman.2017.10.008

Farmaki, A. (2017). The tourism and peace nexus. *Tourism Management, 59*, 528–540. https://doi.org/10.1016/j.tourman.2016.09.012

Felsenstein, D., & Freeman, D. (2001). Estimating the impacts of cross-border competition: The case of gambling in Israel and Egypt. *Tourism Management, 22*(5), 511–521. https://doi.org/10.1016/S0261-5177(01)00006-1

Fujita, M., & Krugman, P. R. (2003). The new economic geography: Past, present and the future. *Papers in Regional Science, 83*(1), 139–164. https://doi.org/10.1007/s10110-003-0180-0

Garín-Muñoz, T., & Moral-Rincón, M. J. (2012). *Outbound Tourism Holiday behavior of Spanish residents: Choice of destination*. Available at SSRN: https://ssrn.com/abstract=2148910 or https://doi.org/10.2139/ssrn.2148910

Garrow, L. (2010). *Discrete choice modelling and air travel demand: Theory and applications*. Ashgate.

Gelbman, A., & Timothy, D. (2019). Differential tourism zones on the western Canada–US border. *Current Issues in Tourism, 22*(6), 623–682. https://doi.org/10.1080/13683500.2017.1304364

Gelbman, A., & Timothy, D. J. (2011). Border complexity, tourism and international exclaves. *Annals of Tourism Research , 38*(1), 110–131. https://doi.org/10.1016/j.annals.2010.06.002

Gobierno Provincial Alicante (Regional Government of Alicante). (2018). *Tourismo por Provincias: Establecimientos Hoteleros*. Retrieved September 9, 2018, from http://documentacion.diputacionalicante.es/ra_turismo.asp

Hall, C. M. (2015). On the mobility of tourism mobilities. *Current Issues in Tourism, 18*(1), 7–10. https://doi.org/10.1080/13683500.2014.971719

Hazledine, T., & Collins, S. (2011). Paying the pilot? The economics of subsidising international air travel to small remote island nations with large diaspora. *Journal of Air Transport Management, 17*(3), 187–194. https://doi.org/10.1016/j.jairtraman.2010.12.008

Hasan-Kerr, T. L. (2018). *Crossing the Ceuta border (From Morocco to Ceuta, Spain by land)*. Online article in Gomadnomad.com. Retrieved June 19, 2019, from https://gomadnomad.com/2018/10/15/crossing-the-ceuta-border/

Hotel Price Index. (2018). *Cuanto pagaron los espanoles para alojarse al extranjero?* Retrieved September 9, 2018, from https://hpi.hotels.com/es-2017/cuanto-pagaron-los-espanoles-por-alojarse-en-el-extranjero/ (in Spanish)

Instituto Nacional de Estadisticas (INE). (2018). *Estadisticas*. Retrieved September 9, 2018, from http://estadisticas.tourspain.es/es-es/estadisticas/fichadecoyuntura/paginas/default.aspx (in Spanish)

Kawulich, B. B. (2005). Participant observation as a data collection method. *Forum: Qualitative Social Research*, *6*(2), Art. 43, http://nbn-resolving.de/urn:nbn:de:0114-fqs0502430.

Koo, T., & Papatheodorou, A. (2017). Spatial evolution of airports: A new geographical economics perspective. In J. Bitzan & J. Peoples (Eds.), *Airport economics – Advances in airline economics series* (Vol. 6, pp. 237–261). Emerald Group Publishing.

Kouwenhoven, M. (2008). *The role of accessibility in passengers choice of airports*. Discussion paper No. 2008-14 prepared for the Round Table of on Airline Competition (October), Systems of Airports and Intermodal Connections, OECD/ITF, Paris, France, October 2–3.

Lagiewski, R., & Revelas, D. (2004). *Challenges in cross-border tourism regions*. Retrieved December 30, 2018, from http://scholarworks.rit.edu/other/551

Liasidou, S. (2013). Decision-making for tourism destinations: Airline strategy influences. *Tourism Geographies*, *15*(3), 511–528. https://doi.org/10.1080/14616688.2012.726266

Lieshout, R. (2012). Measuring the size of an airport's catchment area. *Journal of Transport Geography* , *25*, 27–34. https://doi.org/10.1016/j.jtrangeo.2012.07.004

Litvin, S. (1998). Tourism: The world's peace industry? *Journal of Travel Research*, *37*(1), 63–66. https://doi.org/10.1177/004728759803700108

McNabb, D. E. (2010). *Case research in public management*. M.E. Sharpe Inc.

Office National Des Aéroports (ONDA). (2018). *Statistics*. Retrieved September 9, 2018, from www.onda.ma (in Spanish)

Paliska, D., Drobne, S., Borruso, G., Gardina, M., & Fabjan, D. (2016). Passengers' airport choice and airport's catchment area analysis in cross-border Upper Adriatic multi-airport region. *Journal of Air Transport Management*, *57*, 143–154. https://doi.org/10.1016/j.jairtraman.2016.07.011

Papatheodorou, A. (2002). Civil Aviation Regimes and Leisure Tourism in Europe. *Journal of Air Transport Management*, *8*(6), 381–388. https://doi.org/10.1016/S0969-6997(02)00019-4

Pappas, N., & Papatheodorou, A. (2017). Tourism and the refugee crisis in Greece: Perceptions and decision-making of accommodation providers. *Tourism Management*, *63*, 31–41. https://doi.org/10.1016/j.tourman.2017.06.005

Poulaki, I., & Papatheodorou, A. (2016). Open Skies. *New Vistas Journal*, *2*(1), 16–20.

Poulaki, I., Papatheodorou, A., Kitrinou, E., & Panagiotopoulos, A. (2017). Flying beyond borders: Intermodal considerations to improve accessibility of Aegean Sea Islands, Greece using discrete choice analysis. *Journal of Air Transport Studies*, *8*(2), 94–103. https://doi.org/10.38008/jats.v8i2.34

Poulaki, I., Papatheodorou, A., & Stergiou, D. (2013). Expanding cross border airport catchment area using intermodality: The case of Adnan Menderes International Airport of Izmir. *Tourism Today Fall*, *2013*, 48–60.

Prideaux, B., & Kim, S. (2018). Protocolsas a strategy to reduce travel barriers between countries experiencing or have recently experienced serious political difficulties. *Tourism Recreation Research*, *43*(2), 197–208. https://doi.org/10.1080/02508281.2017.1390634

Procesa. (2016). *Anuarios Estadisticos – Turismo*. Retrieved April 25, 2018, from http://www.procesa.es/attachments/category/51/Ct2016An_C13_Turismo.pdf

Prokkola, E. (2007). Cross-border regionalization and tourism development at the Swedish–Finnish border: Destination Arctic Circle. *Scandinavian Journal of Hospitality and Tourism*, *7*(2), 120–138. https://doi.org/10.1080/15022250701226022

Puerto de Algeciras. (2018). *Statistics of Ferry traffic Algeciras-Ceuta-Algeciras*. Ministerio de Fomento, Gobierno de España. Retrieved September 9, 2018, from http://www.apba.es/estadisticas (in Spanish)

Quandt, R. E. (1970). Introduction to the analysis of travel demand. In R. E. Quandt (Ed.), *The demand for travel: Theory and measurement* (pp. 1–15). Heath Lexington Books.

Rijk, T. (2018). *The politics of (im)mobility ant the border between Ceuta and Morocco* [Master Thesis in Cultural Anthropology]. University of Utrecht (pp. 16–18).

Robinson, G. W. S. (1959). Exclaves. *Annals of the Association of American Geographers*, *49*(3), 283–295. https://doi.org/10.1111/j.1467-8306.1959.tb01614.x

Simandan, D. (2016). Proximity, subjectivity, and space: Rethinking distance in human geography. *Geoforum, 75*, 249–252. https://doi.org/10.1016/j.geoforum.2016.07.018

Sofield, T. (2006). Border tourism and border communities: An overview. *Tourism Geographies, 8*(2), 102–121. https://doi.org/10.1080/14616680600585489

Spasojevic, B., Lohmann, G., & Scott, N. (2018). Air transport and tourism systematic literature review (2000–2014). *Current Issues in Tourism, 21*(9), 975–997. https://doi.org/10.1080/13683500.2017.1334762

Stergiou, A. (2016). Turkey–Cyprus–Israel relations and the Cyprus conflict. *Journal of Balkan and near Eastern Studies, 18*(4), 375–392. https://doi.org/10.1080/19448953.2016.1195994

Stoffelen, A., & Vanneste, D. (2018). The role of history and identity discourses in cross-border tourism destination development: A Vogtland case study. *Journal of Destination Marketing & Management, 8*, 204–213.

Suzuki, S., & Inohara, T. (2015). Mathematical definitions of enclave and exclave, and applications. *Applied Mathematics and Computation, 268*, 728–742. https://doi.org/10.1016/j.amc.2015.06.114

Timothy, D. J. (1995). Political boundaries and tourism: Borders as tourist attractions. *Tourism Management, 16*(7), 525–532. https://doi.org/10.1016/0261-5177(95)00070-5

Timothy, D. J. (1996). Small and isolated: The politics of tourism in international exclaves. *Acta Turistica, 8*(2), 102–121.

Timothy, D. J. (2000). Borderlands: An unlikely tourist destination? *IBRU Boundary and Security Bulletin, 8*(1), 57–65.

Unión Generál de Trabajadores. 2018. *Evolution of remuneration of teachers in public education*. Retrieved December 30, 2018, from https://educacion.fespugtclm.es/wp-content/uploads/2018/05/2017-2010-Comparativa-FeSP-retribuciones-docentes-CCAA%C2%B4s.pdf (in Spanish)

Vinokurov, E. (2007). *The theory of enclaves*. Retrieved April 26, 2018, from https://mpra.ub.uni-muenchen.de/20936/1/MPRA_paper_20936.pdf

**Appendix 1 – Details of travel scenarios**

| City | Destination airport | Origin Airport | a/p access time (hrs-roundtrip) | Travel time (hrs-roundtrip) | a/p access cost (€-roundtrip) | Travel cost (€-roundtrip) | Airline | Stopover (go) | Stopover (return) | **Total time (hrs) access + travel** | Total cost (€) access + travel |
|---|---|---|---|---|---|---|---|---|---|---|---|
| Brussels | BRU | ENG | 2 | 6 | 40 | 57 | Ryanair | | | 8 | 97 |
| | BRU | AGP | 6 | 6 | 100 | 75 | Brussels | | | 12 | 175 |
| Amsterdam | AMS | AGP | 6 | 6 | 100 | 115 | Transavia | | | 12 | 215 |
| | AMS | TNG | 2 | 6 | 40 | 1 83 | Air Arabia Maroc | | | 8 | 223 |
| Paris | BVA | TNG | 2 | 5:20 | 40 | 69 | Ryanair | | | 7:20 | 109 |
| | CDG | AGP | 6 | 5:20 | 100 | 186 | Air Europa | | | 11:20 | 286 |
| London | STN | AGP | 6 | 6 | 100 | 41 | Monarch + Ryanair | | | 12 | 141 |
| | LGW | TNG | 2 | 6 | 40 | 1870 | Air Arabia Maroc | | | 8 | 1910 |
| New York | JFK | TNG | 2 | 25:30 | 40 | 619 | Iberia | MAD | MAD | 27:30 | 659 |
| | JFK | AGP | 6 | 21 | 100 | 466 | KLM | AMS | direct | 27 | 566 |
| Lisbon | LIS | TNG | 2 | 3 | 40 | 157 | TAP Portugal | | | 5 | 197 |
| | LIS | AGP | 6 | 3:10 | 100 | 235 | TAP Portugal | | | 9:10 | 335 |
| Madrid | MAD | AGP | 6 | 2:30 | 100 | 64 | Iberia | | | 8:30 | 164 |
| | MAD | TNG | 2 | 2:40 | 40 | 129 | Iberia | | | 4:40 | 169 |
| Barcelona | BCN | TNG | 2 | 3:30 | 40 | 90 | Vueling | | | 5:30 | 130 |
| | BCN | AGP | 6 | 3:30 | 100 | 40 | Vueling | | | 9:30 | 140 |
| | FRA | AGP | 6 | 6 | 100 | 89 | Lufthansa | | | 12 | 189 |
| Frankfurt | HNN | TNG | 2 | 6:20 | 40 | 56 | Ryanair | | | 8:2.0 | 96 |
| Buenos | EZE | TNG | 2 | 42 | 40 | 1326 | Iberia | MAD | MAD | 44 | 1366 |
| Aires | EZE | AGP | 6 | 43:40 | 100 | 1107 | KLM | AMS | AMS | 49:40 | 1207 |

Sources: www.skyscanner.com,www.rome2rio.com,www.avanzabus.com,www.frs.es,www.balearia.com.www.transmediterra.es (all accessed July 10th 2017), processed by authors.
Note: Regarding airport codes AGP stands for Malaga; TNG for Tangier; BRU for Brussels; AMS for Amsterdam Schiphol; BVA for Paris Beauvais; CDG for Paris Charles de Gaulle; STN for London Stansted; LGW for London Gatwick; JFK for New York JFK; LIS for Lisbon; MAD for Madrid; BCN for Barcelona ; FRA for Frankfurt; HNN for Hahn; and EZE for Buenos Aires.

# Labour migration and tourism mobilities: Time to bring sustainability into the debate

Noel B. Salazar

**ABSTRACT**

Migration and tourism are interconnected forms of human mobility, similar but different. It is impossible to draw neat boundaries around the two because they constantly intersect, sometimes within one and the same individual. Tourism and migration often fuel each other, thereby raising two interesting questions that are rarely asked, namely 'What would tourism be without migration?' and 'What would migration be without tourism?'. In significant ways, sustainability is directly related to the 'mobility' aspects of tourism-related labour. However, in tourism studies, research on mobility often focuses merely on tourist movements. In general, there has been little detailed examination of the mechanisms that comprise and (re)produce the border-crossing movements of tourism labourers. If there is little attention to tourism-related labour mobilities, considerations related to the sustainability implications of worker mobility are highlighted even less often. Indeed, despite the anthropocentric focus of sustainability as a concept, surprisingly little attention has been devoted to the aspect of social sustainability. As the COVID-19 crisis shows, when it comes to the nexus between migration and tourism, the current and future challenges are huge.

**摘要**

迁徙和旅游是相互联系的人类流动形式, 相似但不同。要在两者之间划清界限是不可能的, 因为它们时常交叉, 有时可归为一类, 有时一个人既迁徙又旅游。旅游和迁徙经常相互推动, 因此提出了两个很少被问及的有趣问题, 即"如果没有迁徙, 旅游会怎么样?"以及"如果没有旅游, 迁徙又会怎么样?"在很大程度上, 可持续性与旅游劳工的"流动性"直接相关。然而, 在旅游研究中, 对流动性的研究往往只关注于游客的流动。总体而言, 对于组成和(重新)生成旅游劳工跨境流动的机制, 几乎尚未进行详细的考察过。如果很少注意与旅游有关的劳工流动, 那么就更不经常凸显与工人流动可持续性有关的考量。事实上, 尽管可持续发展是一个以人类为中心的概念, 但令人惊讶的是, 很少有人关注社会可持续性层面。新冠肺炎危机表明, 就移徙和旅游之间的关系而言, 当前和未来的挑战都是巨大的。

Have you ever wondered, when you board a means of transport, where everybody else is going and for which reasons? I have. In fact, I try to ask fellow passengers as soon as the opportunity arises. Particularly planes can contain an incredible mix of travellers. The unprecedented degree of mobility in contemporary society—when not hindered by unforeseen external factors such as COVID-19—is believed to challenge traditional values, including people's attachment to places.[1] So much of social theory is predicated on the thesis of authenticity versus alienation that 'it has been hard to imagine a multicentred world in which movement and mobility play as much of a constituting role in society as more traditionally place-based notions of settlement, territory and community identity' (Williams & Van Patten, 2006, p. 33). While scholars traditionally tended to ignore or regard boundary-crossing journeys as deviations from normative place-bound communities, cultural homogeneity, and social integration, discourses of globalization and cosmopolitanism (that became dominant since the end of the Cold War) shifted the pendulum in the opposite direction (Salazar, 2013a).

People across the globe have long been interconnected; populations often have been mobile; and their identities have long been fluid, multiple, and contextualized. My own family and personal history are deeply marked by multiple mobilities and border crossings. In the early 1970s, my Flemish mother travelled as a young tourist to the Costa del Sol (or 'Sun Coast'), in the south of Spain. Formerly made up of a series of small fishing settlements, the region was converted in the 1960s into a world-renowned beach tourism destination. Like many others living in the *pueblos blancos* ('white villages') a short distance inland in the mountains running down to the coast, my father found employment in the hospitality sector and worked as a hotel receptionist at the time he met my mother. This encounter, in turn, led my Spanish father to migrate to Bruges, Belgium, another famous tourism hotspot. I was born in between, in France, and spent most of my formative years shuttling between Belgium and Spain.

So it is not all that surprising that I ended up doing research on the interconnections between various forms of human border-crossing mobilities (Salazar, 2018). In 2006, for instance, I had the opportunity to visit the remote island of Sumba in eastern Indonesia. At the time, I was conducting ethnographic fieldwork on tourism on the island of Java. The trip to Sumba had been carefully planned because there were only two flights per week with a small propeller plane. Nuri, an old-time Indonesian friend with whom I had studied anthropology in Belgium and who was a native from Sumba, gave an ethnographic tour of the island. On the day of departure, I experienced first-hand what it meant to be 'involuntary immobile', stuck on an island. The Tambolaka airfield was full of agitated people because the flight had been cancelled and nobody could tell when the next plane would come. I was among only a handful of foreigners, along with a French film crew and some lost tourists, who were all annoyed with the fact that their travel itineraries had been heavily disturbed.

Sumba is one of the poorer islands of Indonesia. Most people at the airfield were Sumbanese women who work abroad as migrant labourers, including in the hospitality industry. They had visited relatives in their native villages. For these poor women, the cancellation of the flight had potentially catastrophic consequences. Arriving a couple of days late for work in Malaysia or Singapore could mean the termination of their contract. In other words, even if we were all to take the same plane, their freedom of movement was structurally much more limited than the tourists and others present (Abram et al., 2017).

While mobility was not the focus of my research back then, human movement was evident all around me (Salazar, 2011a). I witnessed, for example, the seasonal flows of tourists and how local tourism service providers were looking for short-term employment abroad during the low season—work on cruise ships or in large hotels—Europe or the United States being popular options. In 2010, I attended the Annual Meeting of the Society for Economic Anthropology in Tampa, Florida. The topic of the conference was 'Contested Economies: Global Tourism and Cultural Heritage'. I therefore thought it fitting to pay a visit to the Port of Tampa, a well-known cruise ship terminal. I witnessed the docking of the MS Carnival Inspiration, a fourteen-deck vessel with the capacity to carry 3,450 passengers and a crew of 920 people. I was particularly interested in the latter population. As passengers were disembarking, I observed many crew members hurrying to the local bars near the cruise port. Outside Hooters, I met several Indonesian and other Asian men. They were not really interested in what was going on inside the bar but wanted to capture the free Wi-Fi signal to communicate with their families back home.

A year later, I was reminded of this encounter while I was interviewing a Romanian woman who founded *Miami*, a school for cruise ship training in Yogyakarta, Indonesia, and who was now running a cruise crew recruitment agency. She herself had worked on a Carnival cruise ship, where she met her future Indonesian husband. She confirmed that many people she recruited were young men, including some of the tour guides I had previously studied (Salazar, 2010). Their main motivation was economic (although they also liked the travel aspect of the job). Because of their limited command of English, most of them ended up doing jobs on board that required little contact with passengers.

This anecdotical evidence serves to illustrate how migration and tourism are interconnected, similar but different forms of mobility. It is impossible to draw neat boundaries around the two because they constantly intersect (Hall & Williams, 2002), sometimes within one and the same person. Many would-be migrants, for example, enter countries on tourist visas. In other instances, tourism and migration become much enmeshed in a certain location. Just think of places such as the islands in the Mediterranean that serve both as tourist destinations and migrant stop-overs (or temporary destinations, in case people get stuck). Moreover, the housing of asylum seekers in Europe usually is organized in accommodations that are often an integral part of the tourism infrastructure (guesthouses, pensions, hostels, and hotels). Because of this practice, tourism spaces are linked with spaces of (forced) migration in various ways.

The connection between tourism and migration is much more intimate than that. At the beginning of the new millennium, Alan Williams and Michael Hall (2002) identified three different mechanisms that link tourism with migration: (1) the tourism-led migration hypothesis (tourism leads to migration, both for employment and consumption); (2) the migration-led tourism hypothesis (migration leads to tourism visiting friends and relatives); and (3) a mutually reinforcing mechanism that accounts for the bidirectional flows between tourism and migration. Undoubtedly, both forms of mobility fuel each other, thereby raising two interesting questions. The first one is 'What would tourism be without migration?' Well, global tourism organized at the massive scale we have known until now is impossible without the availability of both a numerically substantial and functionally flexible workforce of migrant workers. The labour-intensive and casual nature of tourism accentuates the role of various economic sectors, most of which are loosely grouped under the term 'hospitality industry' (particularly hotels and restaurants), as a magnet for migrant labour. Given the increasing transnational nature of tourism, it is hardly surprising that tourism employment is 'cosmopolitan' in nature (Salazar, 2017), although there are clear differences related to age, ethnicity, and gender (Bianchi, 2000). A tourism workforce that is mobile offers 'a solution to labour shortages where the local workforce is not willing to engage in low pay, low status and seasonal employment' (Janta et al., 2011, p. 1323).

This brings us to the second question: 'What would migration be without tourism?' Not only does tourism provide migrant labour, many migrants are also partly inspired by tourism-related imaginaries to plan their migratory itineraries (Salazar, 2010). In the cultural logics of migration, imaginaries play a predominant role in envisioning the green grass on the other side (Salazar, 2011b). Not surprisingly, many of these imaginaries are circulated through tourism and its various stakeholders. Moreover, mobile lifestyles evolve not just to explore economic opportunities not available locally but also to pursue particular types of culturally and socially desirable livelihoods, what is better known as 'lifestyle migration' (Benson & Osbaldiston, 2014) or 'lifestyle mobility' (Cohen et al., 2015).

Even if tourism and migration often overlap, in practice tourist and migrant movements are governed by very different regimes of mobility (Salazar & Glick Schiller, 2014). By using mobility as an analytical lens, scholars became more aware of this. It is not because one focuses attention on the 'fluid' aspects of society that societal structure disappears entirely (Salazar, 2016). Barriers to border-crossing movements typically increase after big 'crises' and are accompanied by the counter narrative of securitization. Just think of 9/11 in the United States, the 2015 refugee influx and terrorist attacks in the EU and the 2020 global crisis caused by COVID-19. In the context of the tourism labour market, the differential regimes are partially related to the type of people that are employed. Tourism largely relies on 'low-skilled' workers (having skills that are not highly valued) and workers with few qualifications in general, people belonging to (ethnic) minorities, unemployed youngsters, people who are long-term unemployed, and people (mostly women) who can only commit to part-time jobs because of responsibilities within the family. It are mostly those groups that are hit hardest in times of crisis. The high labour mobility is not only related to the seasonality of tourism activities but also to the high turnover of staff joining and leaving the sector (Bookman, 2006).

Sometimes the differences between tourism and migration regimes of mobility become very visible. From 1999 until 2012, for example, all *Tenaga Kerja Indonesia* (TKI) or '(overseas) Indonesian labour migrants' had to use the special Selapajang Terminal at Soekarno-Hatta International Airport in Jakarta. This terminal was located at a distance from the other airport terminals. Return migrants were welcomed back as *pahlawan devisa* (foreign exchange heroes) because of the remittances they brought with them. The aim of creating a separate terminal for Indonesian migrant workers was to protect them against corrupt money changers or transportation operators. In practice, however, separating these people in a distant terminal only made them more vulnerable.

Given this general context of migration and tourism mobilities, many issues of sustainability arise. In the remainder of this article, then, I want to focus on the questions of sustainability that are directly related to the 'mobility' aspects of tourism labour. I conceptualize mobility here as a complex assemblage of movement, social imaginaries, and experience (Salazar, 2013b). As a sociocultural anthropologist, I am particularly interested in the culturally inflected meanings, values, and impacts of mobilities. Moving beyond the conventional approach to human movement in clearly delineated subfields (e.g. migration and tourism studies), I want to address the normalization of certain boundary-crossing movements and the relations of differential power that are generative of these mobilities, their representations, and their societal significance.

## *Tourism-related labour mobilities*

In tourism, studies of human mobility are often limited to tourist movements. Through their 'Statistics and Tourism Satellite Account Programme', for instance, UNWTO has very detailed data on 'international arrivals' across the globe. Unfortunately, the data on the mobility of workers who make tourism possible are much less detailed. In general, there has been little thorough examination of the mechanisms that comprise and (re)produce the border-crossing movements of tourism-related labourers. As Raoul Bianchi noted already two decades ago, 'There are a few notable exceptions, which encompass a relatively broad spectrum of tourism-related migration, but which for the most part are concerned with the more familiar category of *labour* migration as opposed to transient resort workers who do not fit easily into the category of either tourist or migrant' (2000, p. 112). This is remarkable because waves of temporary migration in the tourism sector are not at all new. Tom Baum (2006) traces examples of vocational mobility in the sector in Europe back to the 13th century.

There are many categories of mobile tourism workers. Doris Anna Carson and colleagues (2014, p. 359), for example, describe the following categories: (1) migrant tourist workers (who have mostly lifestyle-related motivations and repetitively travel between places and jobs); (2) practiced seasonal workers (who combine seasonal tourism work with jobs during off season elsewhere and have relatively predictable cyclical patterns of mobility); (3) working-holiday tourists (who chase temporary work opportunities to earn income while travelling the world); (4) temporary seasonal workers (who have more random work patterns and may engage in seasonal tourism work as a one-off opportunity, usually for economic reasons); (5) hobbyists (who work in tourism to satisfy personal leisure interests); and (6) place-attached seasonal workers (who live in the seasonal workplace or have prior connections to the place through family ties or previous tourist experiences). These categories, which are not necessarily mutually exclusive, emphasize the complex, and often blurred, relationships between tourism, work, and labour migration (Bianchi, 2000).

## The social sustainability of tourism-related labour mobilities

If there is little attention to tourism-related labour mobilities, 'considerations related to the sustainability implications of worker mobility and migration are highlighted even less often' (Baum et al., 2016, p. 7). The concept of sustainable development was developed alongside the heightened consciousness that ecological devastation, together with a 'retreat from social concerns' (e.g. poverty), is untenable (WCED, 1987, p. xi). Even if the original sustainable development concept included a clear social focus, for the first decades 'this human dimension has been neglected amidst abbreviated references to sustainability that focused on bio-physical environmental issues, or was subsumed within a discourse that conflated "development" and "economic growth"' (Vallance et al., 2011). The social aspects started receiving far more global attention through the UN's Millennium Development Goals (2001-2015) and Sustainable Development Goals (2016-2030), the latter now being criticized by some for being too weak on environmental sustainability.

As Nicola Dempsey and colleagues note, 'Social sustainability is a wide-ranging multi-dimensional concept, with the underlying question "what are the social goals of sustainable development?", which is open to a multitude of answers, with no consensus on how these goals are defined' (Nicola et al., 2011, p. 290). In 2005, the United Nations Environment Programme and the United Nations World Tourism Organization jointly released a guide for policy makers, entitled *Making Tourism more Sustainable* (UNEP/UNWTO, 2005). This guide describes social sustainability as 'respecting human rights and equal opportunities for all in society', requiring 'an equitable distribution of benefits, with a focus on alleviating poverty' (UNEP/UNWTO, 2005, p. 9). Important is the 'emphasis on local communities, maintaining and strengthening their life support systems, recognizing and respecting different cultures and avoiding any form of exploitation' (UNEP/UNWTO, 2005, p. 9).

In terms of the nexus between migration and tourism, the current and future challenges are huge. For the Asia-Pacific region, for example, the diametrically opposed mobility trends of tourists and migrants have been pointed out (Robinson et al., 2014). According to Richard Robinson et al., many tourists in this region are attracted by 'isolated' (remote) destinations. However, the global trend towards urbanization draws the workforce to metropolitan areas. These two trends 'create major challenges for the tourism industry, which faces a growing demand for labour in both areas where there is growing labour supply but without skills (untrained agricultural workers migrating to core megacities) and where there is a shrinking labour supply of all employees (pleasure periphery)' (Robinson et al., 2014, p. 806). It is clear, however, that in a post-COVID-19 world peripheral destinations will have a hard time because the mobilities of both tourists and tourist workers will be more complicated. More central locations, on the other hand, may be avoided because they concentrate too many people.

Worryingly, sustainable tourism planning and development in general is threatened by quickly spreading neoliberal ideas of resilience. Resilience in general refers to the ability to resist and recover from aversity or disaster. When resilience is missing, crises will be suffered as unsettling; when resilience is robust, a crisis will be experienced more smoothly. Alan Lew notes how the planning of resilience has emerged as an alternative to the sustainable development paradigm. The difference between the two is that 'sustainability mitigates or prevents change by maintaining resources above a normative safe level, whereas resilience adapts to change by attempting to build capacity to return to a desired state following both anticipated and unanticipated disruptions' (Lew, 2014, p. 14). Sustainability emphasizes 'aspirational goals associated with the careful use of resources and ensuring provision for future generations' while resilience is 'pragmatic and inclusive of a range of responses that may or may not align with sustainability principles' (Espiner et al., 2017, p. 11). The neoliberal framing of resilience is a way to maintain the status quo: it wants people to be prepared against any aversity or disaster that come their way instead of working towards systemic change.

Since the late 1990s, many stakeholders involved in tourism have addressed issues of sustainability. More recently, they have started looking into capacity development (resilience), so that the tourism sector is ready to respond to various types of disaster. However, as Stephen Espiner and colleagues argue, 'Few ... have demonstrated a capacity to adapt to incremental threats to their longevity (sustainability). In the tourism literature, there is considerable emphasis on resilience to the immediate challenges (local impacts, disasters or financial shocks, for instance), yet there is merit in conceptualizing resilience as a dynamic long-term state, where there are obvious parallels with the sustainability concept.' (Espiner et al., 2017, p. 2). While destinations may manage to be resilient without being sustainable, the opposite is not workable. In other words, in order to be sustainable, they also need to be resilient. The post-COVID-19 period will offer a natural laboratory to study and compare tourism destinations in which principles of sustainability and/or resilience are applied in different ways. For one, how will destinations deal with the predicted loss of (migrant) jobs, particularly in tourism?

The question what constitutes sustainable tourism becomes even more complex when we take labour migration into account. Even if the absolute numbers of labour mobilities may be small, they can raise delicate issues locally that need to be considered when assessing the value of tourism as a local development tool (Carson et al., 2014). Examples include the lack of local integration of 'enclave' labour communities, regular economic and knowledge leakages due to high labour turnover, and the lack of long-term opportunities for a more stable and locally based workforce (Tuulentie & Heimtun, 2014). Indeed, skilled migrant workers, particularly those who settle, can actually 'displace' local workers (Schellhorn, 2010). The problem is that there is not one global model of sustainability that can be easily applied in all tourism contexts (or that too many so-called experts have unsuccessfully tried this around the world in the past). How to make tourism and its labour-related migration (more) sustainable is dependent on many context-specific variables. In other words, every destination may need a strategy with different accents.

Another issue is that increasing migrant mobility and the presence of greater numbers of migrants in a tourism destination also carries consequences for tourism in terms of the product and experience offered to tourists and the imaginary they cultivate of a destination (Salazar, 2012). Migrants employed in 'front-of-house' positions interact with the local community as well as domestic and international tourists (Janta et al., 2011, p. 1339). Interestingly, those workers play the role of 'host' for international tourists. In the case of new arrivals, migrants thus play a double role: they are 'the migrant-guest', but they are also 'migrant- host' to those they serve in tourism. A similar blurring of categories occurs in the case of tourists (often of the 'backpacker' type) who stay longer at a destination and temporarily work in tourism as a way to fund their travels. In other words, the many cross-overs between migration and tourism also put into question the often-used host-guest paradigm. More importantly, perhaps, they make us reflect on the various meanings of 'hospitality' (both within tourism and beyond).

## Conclusion

In-depth research on the situated articulations between culturally inflected regimes of movement, mobility representations, and ideas about tourism and migration is a fruitful way for analysing the dynamic tension between ongoing processes of mobility and fixity. It is necessary, for instance, to understand tourism and migration within the context, and as counter-veiling forces, of a broader reading of globalisation and of uneven and unequal development (Mowforth & Munt, 2016). While the frontiers of tourism are dismantled through the liberalisation of the service sector, the *cordon sanitaire* against migration to the developed economic core is tightened, even though migrants' remittances are, in some cases, more significant than earnings from tourism. At the same time, crises such as the one caused by COVID-19 can quickly unsettle these mechanisms and processes. By juxtaposing migration and tourism mobilities we emphasise the framework of global unevenness and inequality in which all forms of human movement must be contextualised and are conditioned. Whilst many global migrants are relocating for work, family and/or lifestyle, many others have very little choice of when and where to go as they flee their homes as equally diversely displaced refugees. Climate change and future crises and disasters will only exacerbate forms of forced mobility.

The connections between migration and tourism (and vice versa) have become increasingly important 'because of significant shifts in the role that migration plays in the international political and economic order and the consequent shifting definitions of what constitutes a migrant (and tourist) in the twenty-first century' (Lovelock & Leopold, 2011, p. 139). However, although increasingly attracting research attention, there are aspects that remain poorly investigated. How, for one, can we ensure that regimes of mobility do not disrupt, but rather, facilitate and support the mobilities and livelihoods of local populations, especially the poor and vulnerable?

An issue that urgently requires our attention and research is the impact of human mobility and everything that comes with it. This impact is becoming continuously greater because of the number and frequency of human movements across the planet and because of the (polluting) means of transport used. In general, the whole (environmental) sustainability argument seems to have very limited impact on how most people imagine, experience, and value translocal mobilities. This is a huge challenge, one that will only grow in the future. Regrettably, the COVID-19 crisis has pushed the problem of climate change somewhat out of the spotlight (despite the fact that climate change exacerbates the risk of pandemics), but it has made more people reflect on how and why we move across the planet and how unsustainable this all is.

Within tourism, too, there is much thinking about how the future of the sector should look like. It is of crucial importance to include considerations regarding the tourism-migration nexus. Sustainable tourism workforce mobility requires 'coordinated and integrated policy and planning approaches at the local, regional, national and even transnational levels' (Baum et al., 2016, p. 8). Moreover, 'any discussion about career management and turnover in tourism organizations in specific locations must be seen within the contemporary understandings of vocational mobility and global migration patterns'(Baum et al., 2016, p. 9). As I pointed out earlier, unfortunately there is not an easy one-type-fits-all solution. Even best practices or successful examples are not to be blindly copied (although they may certainly give inspiration). Because of the particularities of each destination or setting in which tourism and its related labour migration are embedded, we need detailed studies and the involvement of as many stakeholders as possible to work towards 'sustainable' sustainable tourism development.

Humans have always been mobile. People are on the move in multiple directions. There is no single model or grand theory that can explain the complexity of human mobility, certainly not on a global scale. This implies that we need to pay much more attention, both in research and praxis, to alter-native (particularly non-western) understandings of mobilities that are commonly labelled as 'tourism' and 'migration' (Adams, 2020). The mobility experience that many people aim for is less about regularly moving back and forth between 'home' and 'elsewhere', but encompasses a broader understanding of movement, which includes a variety of locations that emerge and/or are abandoned organically in an individual's life trajectory. Therefore, a semiotic analysis of mobility seems in place, a non-binary inquiry into the meanings of home, of being away, and the consequences certain articulations of home, mobility, and the wider world might have for the constructions of self and place (Teampău & Van Assche, 2009). Or, in the words of anthropologist Tim Ingold (2011, p. 12), 'We need a different understanding of movement: … not the *trans-port* (carrying across) of completed being, but the *pro-duction* (bringing forth) of perpetual becoming'.

## Note

1. This article is based on a keynote speech I gave for the 'Nexus of Migration and Tourism: Creating Social Sustainability Symposium' at VNU University of Social Sciences and Humanities, Hanoi, Vietnam in September 2018. I thank Long Hong Pham and his team at VNU for the invitation and Jaeyeon Choe from Bournemouth University in the UK for facilitating the contact, the intellectually stimulating exchanges and the pleasant company. As the revision of the article happened in the middle of the 2020 coronavirus pandemic, reflections relating to COVID-19 were added at that stage.

## Disclosure statement

No potential conflict of interest was reported by the author(s).

## ORCID

Noel B. Salazar http://orcid.org/0000-0002-8346-2977

## References

Abram, S., Feldman Bianco, B., Khosravi, S., Salazar, N. B., & de Genova, N. (2017). The free movement of people around the world would be Utopian. *Identities: Global Studies in Culture and Power, 24*(2), 123–155. https://doi.org/10.1080/1070289X.2016.1142879

Adams, K. M. (2020). What western tourism concepts obscure: intersections of migration and tourism in Indonesia. *Tourism Geographies*, 1–26. https://doi.org/10.1080/14616688.2020.1765010

Baum, T. (2006). *Human resource management for tourism, hospitality and leisure: An international perspective*. International Thomson.

Baum, T., Cheung, C., Kong, H., Kralj, A., Mooney, S., Nguyễn Thị Thanh, H., Ramachandran, S., Dropulić Ružić, M., & Siow, M. (2016). Sustainability and the tourism and hospitality workforce: A thematic analysis. *Sustainability, 8*(8), 809. https://doi.org/10.3390/su8080809

Benson, M., & Osbaldiston, N. (Eds.). (2014). *Understanding lifestyle migration: Theoretical approaches to migration and the quest for a better way of life*. Palgrave Macmillan.

Bianchi, R. V. (2000). Migrant tourist workers: Exploring the contact zones of post-industrial tourism. *Current Issues in Tourism, 3*(2), 107–137. https://doi.org/10.1080/13683500008667869

Bookman, M. Z. (2006). *Tourists, migrants & refugees: Population movements in Third World development*. Lynne Rienner.

Carson, D. A., Carson, D. B., & Lundmark, L. (2014). Tourism and mobilities in sparsely populated areas: Towards a framework and research agenda. *Scandinavian Journal of Hospitality and Tourism, 14*(4), 353–366. https://doi.org/10.1080/15022250.2014.967999

Cohen, S. A., Duncan, T., & Thulemark, M. (2015). Lifestyle mobilities: The crossroads of travel, leisure and migration. *Mobilities, 10*(1), 155–172. https://doi.org/10.1080/17450101.2013.826481

Espiner, S., Orchiston, C., & Higham, J. (2017). Resilience and sustainability: A complementary relationship? Towards a practical conceptual model for the sustainability–resilience nexus in tourism. *Journal of Sustainable Tourism, 25*(10), 1385–1400. https://doi.org/10.1080/09669582.2017.1281929

Hall, C. M., & Williams, A. M. (Eds.). (2002). *Tourism and migration: New relationships between production and consumption*. Kluwer.

Ingold, T. (2011). *Being alive: Essays on movement, knowledge and description*. Routledge.

Janta, H., Brown, L., Lugosi, P., & Ladkin, A. (2011). Migrant relationships and tourism employment. *Annals of Tourism Research, 38*(4), 1322–1343. https://doi.org/10.1016/j.annals.2011.03.004

Lew, A. A. (2014). Scale, change and resilience in community tourism planning. *Tourism Geographies, 16*(1), 14–22. https://doi.org/10.1080/14616688.2013.864325

Lovelock, K., & Leopold, T. (2011). The political economy of temporary migration: Seasonal workers, tourists and sustaining New Zealand's labour force. In J. Mosedale (Ed.), *Political economy of tourism* (pp. 139–156). Routledge.

Mowforth, M., & Munt, I. (2016). *Tourism and sustainability: Development, globalisation and new tourism in the Third World* (4th ed.). Routledge.

Nicola, D., Glen, B., Sinéad, P., & Brown, C. (2011). The social dimension of sustainable development: Defining urban social sustainability. *Sustainable Development, 19*(5), 289–300. https://doi.org/10.1002/sd.417

Robinson, R. N. S., Ritchie, B. W., Kralj, A., Solnet, D. J., Baum, T., & Ford, R. C. (2014). An Asia-Pacific core–periphery futures paradox: Divergent worker and tourist mobilities. *Journal of Travel Research, 53*(6), 805–818. https://doi.org/10.1177/0047287513513164

Salazar, N. B. (2010). *Envisioning Eden: Mobilizing imaginaries in tourism and beyond*. Berghahn.

Salazar, N. B. (2011a). Grounding mobilities: Rethinking border-crossing tourism and migration. In E. Judd & J. Zhang (Eds.), *Labor migration and social mobility in Asia and Pacific region* (pp. 8–25). Intellectual Property Publishing House.

Salazar, N. B. (2011b). The power of imagination in transnational mobilities. *Identities: Global Studies in Culture and Power, 18*(6), 576–598. https://doi.org/10.1080/1070289X.2011.672859

Salazar, N. B. (2012). The (im)mobility of tourism imaginaries. In M. Smith & G. Richards (Eds.), *The Routledge handbook of cultural tourism* (pp. 34–39). Routledge.

Salazar, N. B. (2013a). Anthropology. In P. Adey, D. Bissell, K. Hannam, P. Merriman, & M. Sheller (Eds.), *The Routledge handbook of mobilities* (pp. 55–63). Routledge.

Salazar, N. B. (2013b). Mobility. In R. J. McGee & R. L. Warms (Eds.), *Theory in social and cultural anthropology* (Vol. 1, pp. 552–553). Sage.

Salazar, N. B. (2016). Conceptual notes on the freedom of movement and bounded mobilities. In M. Gutekunst, A. Hackl, S. Leoncini, J. S. Schwarz, & I. Götz (Eds.), *Bounded mobilities: Ethnographic perspectives on social hierarchies and global inequalities* (pp. 283–289). Transcript.

Salazar, N. B. (2017). The cosmopolitanization of tourism: An afterthought. In R. Shepherd (Ed.), *Cosmopolitanism and tourism: Rethinking theory and practice* (pp. 187–194). Lexington.

Salazar, N. B. (2018). *Momentous mobilities: Anthropological musings on the meanings of travel*. Berghahn.

Salazar, N. B., & Glick Schiller, N. (Eds.). (2014). *Regimes of mobility: Imaginaries and relationalities of power*. Routledge.

Schellhorn, M. (2010). Development for whom? Social justice and the business of ecotourism. *Journal of Sustainable Tourism, 18*(1), 115–135. https://doi.org/10.1080/09669580903367229

Teampău, P., & Van Assche, K. (2009). Migratory marginalities: Making sense of home, self and mobility. *Ethnologia Balkanica, 13*, 147–162.

Tuulentie, S., & Heimtun, B. (2014). New rural residents or working tourists? Place attachment of mobile tourism workers in Finnish Lapland and Northern Norway. *Scandinavian Journal of Hospitality and Tourism, 14*(4), 367–384. https://doi.org/10.1080/15022250.2014.967998

UNEP/UNWTO. (2005). *Making tourism more sustainable: A guide for policy makers*. UNEP/UNWTO.

Vallance, S., Perkins, H. C., & Dixon, J. E. (2011). What is social sustainability? A clarification of concepts. *Geoforum, 42*(3), 342–348. https://doi.org/10.1016/j.geoforum.2011.01.002

WCED. (1987). *Our common future*. Oxford University Press.

Williams, D. R., & Van Patten, S. R. (2006). Home *and* away? Creating identities and sustaining places in a multi-centred world. In N. McIntyre, D. R. Williams, & K. McHugh (Eds.), *Multiple dwelling and tourism: Negotiating place, home, and identity* (pp. 32–50). CABI.

# Index

Page numbers in **bold** refer to tables and those in *italic* refer to figures.


Abreu, J. 143
Accessing Malaga (AGP) airports *144*, 147, *151*
acculturation skills 111
Adams, K. M. 2
Adnan Menderes International Airport 145
AGP/TNG Relative Traffic 2011–2017 *150*
Agriculture & Agri-Food Canada 74
Aguilar Ros, A. 135
Ahmed, S. 84
Airbnb, Booking.com 14
Airbnb rent gap 13
airports and cruise quays 13
Almeida Santos, C. 127
Andalusian Tourism Register 15
Ando, K. 46
Andres, J. 85
anomic cultural relationships 101
association graphs 60
Autonomous Communities 145
Avieli, N. 41

Balanced Outlook 109
Barcelona (BCN) 153
Basok, T. 86
Baum, T. 170
better-off foreigners 14
Binford, L. 86
Biogeme software version 1.8 151
Bowleg, L. 122
Bramley, G. 12
Brexit and terrorist attacks 3
Brown, C. 12
The Budapest Declaration (2002) 36
bunkhouses 85

Carson, D. A. 170
Carson, D. B. 170
Catudal, H. M. 142
Ceuta context 143, 147–9, **148**, *149*
Chan, J. H. 48
cheap direct flights 97
Chinese Canadians (2006) 84
City Master Plan of 2006 18
cohesion and inclusiveness 35
Cole's intersectional analysis (2017) 122
collaborative DataHippo project 15
commodification 37
community members 89
concourse development 103–4
congestion 10
Contract for the Employment in Canada of Commonwealth Caribbean Seasonal Agricultural Workers 86
'core-and-buffer-zone' principle 36–7
cosmopolitan cultures 3
cosmopolitanism 167
craft villages 47
Crenshaw, Kimberle 118, 123
Crouch, G. I. 145
cultural awareness education 111
cultural homogeneity 167
cultural tourism destinations 13
Currid-Halkett, E. 120–1

Davis-Floyd, R. 120
demographic transformation 45–7
Dempsey, N. 12, 171
descriptive statistics 105
destination specificity 152
discourses of globalization 167
discrete choice analysis 155–8
Disneyland 14
displacement 26–8
dummy variables 152

Employment & Social Development Canada 74, 85
*Employment Standards Act* (2000) 81, 86
empowerment 119
Espiner, S. 172
exclaves 142; accessibility, air transport and tourism 145–7

Fageda, X. 143, 145
Farmaki, A. 146

farm-to-table, NOTL 79–81
Feliciano, Jesus Sanchez 88
Ferry Traffic on Algeciras-Ceuta Line 2013–2017 *149*
forced migration 100
Franco, Francisco 16
French obstetricians 121
'fringe stakeholders' 111
Fukukawa, Y. 46

gambling 146
gentrification 11, 12, *19*, 19–21, *20*, 37–8
Geographical Information System (GIS) 40
Goodwin, H. 13
Gotham, K. F. 48
Grimwood, B. S. R. 123
Guarded Mistrust 109

Halal policy 100
Hall, C. M. 58
Hall, Michael 168
Hallyu (Korean Wave) product 100
Hasan-Kerr, T. L. 154
Hebson, G. 122
Hennebry, J. 86
heritage management 35
heritage tourism 76–9, *77*, *78*
Higham, J. 172
High Court Mabo decision (1992) 84
hobbyists 170
Hoi An 34–5, *40*, 40–1; economic performance, tourism sector **43**; expanded conservation zone *42*; international and domestic tourist trends *42*; population by ward (2001–2010) *43*
The Hoi An Center for Monument Management and Preservation (HACMMP) 45
Hoi An City Portal 46, 47
Holgate, J. 122
Holistic Processing of Complexity 60
HomeAway platform 14
Hotel Price Index (2018) 151
Huang, S. 48
human border-crossing mobilities 167
humanized birth *vs.* traditional birth 119–22
human mobility 58
hyper-self-reflexivity 126–7

Iankova, K. 48
"intersectionality" concept 118, 119, 122
intersectional racialization processes 118

Japanese Canadians (1988) 84
JCU-AGP services 153
Jeju island and Yemeni migrants 99–101
Jimenez, J. L. 143
journalists 89

Kernel method 15
Koens, K. 14
Korean ethnonationalism 100

Labour Market Impact Assessment (LMIA) 81–2
*'Labour migration and tourism mobilities: Time to bring sustainability into the debate'* 5
language barriers 87
Lantern Festival 47
Leslie, D. 123
Lew, A. A. 47, 171–2
Liaison Officers 86
lifestyle migrants 2
lifestyle migration, Seville's historic district 25–6
'living heritage approach' 35
local farms, global labour 81–2
*'Local gastronomy, transnational labour: Farm-to-table tourism and migrant agricultural workers in Niagara-on-the-Lake, Canada'* 5
Lommerse, M. 134
*Lonely Planet* guidebook 41
long-term migrants 2
'low-skilled' workers 169
Lundmark, L. 170

Madrid (MAD) 153
Marcus, G. E. 123
McBride, A. 122
McDonald, T. 48
media representations 3
Metcalfe, P. 134
Migrant Agricultural Worker Program 87
migrant labour 3
migrants *vs.* tourists *67*
migrant-tourism phenomenon 97
migrant tourist workers 170
migrant workers 2
migration decisions 58
migration-led tourism hypothesis 168
Milano, C. 14
mobile actors 58
mobility/immobility 58, 143
Mooney, Shelagh 122
Moroccan airports 143
Moroccan passport-holders 154
Muldoon, M. 123
multinomial logit (MNL) model 151, **155**
Murray, I. 16
museumification 37–8, 44–5

Nader, L. 120
The Nation comes First 107–8
negative socio-economic consequences 5
Ng, P. T. 47
Niagara-on-the-Lake (NOTL) 5; farm-to-table 79–81; heritage tourism 76–9, *77*, *78*; labouring bodies 84–5; language of heritage 79; local gastrotourism 76; migrant farm workers 74, 75; regulated invisibility 85–9; tourism and agricultural production 75; visibility/invisibility 82–4
Niagara region map *77*
Niagara's Peach Festival 74
Ni, C. C. 47

non-government organizations (NGOs) 40
normative place-bound communities 167

*Occupational Health and Safety Act* (1990) 86
often-used host-guest paradigm 173
Ontario Ministry of Labour 81
open coding preocess 125
operant subjectivities: attitude changes 111; by cluster **110**; structural changes 111
Operational Program of Ceuta for 2014–2020 147
Orchiston, C. 172
Organization of Indigenous Doctors of the State of Chiapas (OMIECH) 121
Ortisei, touristic hot spot in Dolomites 63–7
Outstanding Universal Values 77–8
overtourism 13, 14, *20*, *21–2*, 21–5

Papatheodorou, A. 145, 146, 149–50
Papp, B. 14
Parnwell, M. 45–7
participant observation 152
*partos humanizados* 120
Patois speakers 86
payment schemes 121
Peach Pickers Picnic *88*
Pearson correlation coefficient 59, 62
pene-exclaves 142
*'Perceptions of and interactions between locals, migrants, and tourists in South Tyrol'* 5
place-attached seasonal workers 170
Postma, A. 14
Poulaki, I. 145, 146, 149–50
power relations 101, 104
Power, S. 12
practiced seasonal workers 170
Preibisch, K. 86
privatisation of public spaces 10
professional midwives 120
protests and mobilization 10
Provincial Statistics Institute 59, 60
P-set 103, 104, **107**
Public Service Obligations (PSOs) 145
public space appropriation 13
'pure' blood nation 100

Qiu, J. 123
Qi, X. 48
Q method: 'clusters of subjectivities' 103; concourse development 103–4; descriptive statistics 105; highest distribution scores **108**; Jeju island and Yemeni migrants 99–101; operant subjectivities, cluster **110**; process methods 103; residents' operant subjectivities 98; statistical and interpretive analysis 105
*'Q method finds anti-refugee sentiments on Yemeni migration to Jeju'* 5
Q-set 103, 104, **106**
qualitative research strategy 59–60
Quandt, R. E. 155
Quang Nam Province 34
quantitative research strategy 59
Quest Community Health Centre (2017) 87

Rapp, R. 123
real estate speculation 10
The Refugee Convention 99
refugee migrants 102
refugee migration 97–9
researchers 89
retail transformation 13
revitalization *19*, 19–21, *20*
Rijk, T. 154
Ritchie, J. R. B. 145

Salazar, N. B. 3, 5, 6
salespeople 89
Schiller, N. G. 3
Seasonal Agricultural Worker Program (SAWP) 74, 82
sea transport 143
second-home-owners 98
self-romanticization 128–31
Seville's historic district: displacement and social unsustainability 26–8; lifestyle migration 25–6; overtourism *20*, 21–2, 21–5; revitalization and gentrification *19*, 19–21, *20*; tourist space 17–19, *18*
Seville's official tourism website 15
The Shades of Summer capstone event 74
short-term migrants 2
Smith, D. 134
Smith, N. 21
social-ecological systems 38
social equity 12
social exchange theory 101, 104
social integration 167
'social polarisation' 34
social sustainability: defined 4; economic and social aspects 12; in heritage towns 35–6; tourism-related labour mobilities 171–3; and WHS management 36
social unsustainability 26–8
socio-economic indicators 43–4
Southern Europe facing Touristification (SET) network 11
South Tyrol's tourism-migrant nexus 60–3, *61*, **62**
Spain and Andalusia, foreigners 16–17
Spanish airports 143
Spanish-speaking workers 86
Spanish Statistics Office 19
Spearman correlation coefficient 59, 62
stakeholders 101
stakeholder subjectivity 102
Stay Grounded declaration 13
Stergiou, D. 145

Sumba island, Indonesia 167–8
Su, M. M. 48
'sun and sand' destinations 11
Sustainable Development Goals (2016–2030) 171

Tangier (TGP) airports *144, 151*
temporary foreign worker (TFW) 81–2
temporary/seasonal migrant workers 2, 170
Tenaga Kerja Indonesia (TKI) 169
Teo, P. 48
Terminal High Altitude Area Defense (THAAD) system 99
Timothy, D. J. 147
tolerant multiculturalism 84
Tomoda, H. 46
'tourism enhancing effect' 34
*Tourism Geographies* 4
tourism indicators 59
tourism-led-migration hypothesis 57, 168
tourism-related housing 10
tourism-related labour mobilities 170
tourist beds in Seville *21*
traditional births 121
Traditional Medicine and Intercultural Development (TMID) 128
"Traditional Mexican Midwifery": ethnomedical tourism 119; geographic distribution of participants **125**; hyper-self-reflexivity 126–7; intersectional axes of analysis 127; Mexico movement 119; national and international scale 121; 'neo-colonial phenomenon' 131–4; neocolonial tourism 118; self-romanticization 128–31; specific methods 123; types of participants **125**
*'"Traditional Mexican Midwifery" tourism excludes indigenous "others" and threatens sustainability'* 5
traditional midwives 120
transnational gentrification process 12, 13
travelers, types of 58
travel scenarios **165**
trust and social responsibility 102–4

Ukrainian Canadians (2008) 84
UNESCO-endorsed spatial planning framework 34
union workers 89
UN's Millennium Development Goals (2001–2015) 171
urban social sustainability 12

Van Wagenen, A. 85
visiting-friends-and-relatives (VFR) 151
voluntary migration 98

Wall, G. 48
'white villages' *(pueblos blancos)* 167
*'Who is the city for? Overtourism, lifestyle migration and social sustainability'* 4
Williams, Alan 168
Williams, A. M. 58
Williams, T. R. 131
Wilson, A. 123
Wine Country Vintners 80
working-holiday tourists 170
World Heritage Convention in 1977 36
World Heritage Sites (WHSs) 4; 'core-and-buffer-zone' principle 36–7; gentrification and museumification 37–8; long-term social sustainability 34–5; and management 36; negative and positive impacts 34; 'social polarisation' 34; 'tourism enhancing effect' 34
The World Tourism Organization 13
Wu, T. C. 47

Xu, K. 48

Yan, G. 127
Yemeni refugee migration 97
Yemenis are Wolves in Sheep's Clothing 108–9
Yudina, O. 123

Zhang, Y. 48
Zukin, S. 12